MAKING AND UNMAKING OF PUGET SOUND

MAKING AND UNMAKING OF PUGET SOUND

Gary C. Howard and Matthew R. Kaser

CRC Press
Taylor & Francis Group
Boca Raton London New York

CRC Press is an imprint of the
Taylor & Francis Group, **an informa** business

Cover Image: NASA.

First edition published 2022
by CRC Press
6000 Broken Sound Parkway NW, Suite 300, Boca Raton, FL 33487-2742

and by CRC Press
2 Park Square, Milton Park, Abingdon, Oxon, OX14 4RN

CRC Press is an imprint of Taylor & Francis Group, LLC

Library of Congress Cataloging-in-Publication Data
A catalog record has been requested for this book

ISBN: 978-1-138-59679-5 (hbk)
ISBN: 978-1-032-20118-4 (pbk)
ISBN: 978-0-429-48743-9 (ebk)

DOI: 10.1201/9780429487439

Typeset in Times
by MPS Limited, Dehradun

Contents

About the Authors

Gary C. Howard recently retired as the Manager of Scientific Editing at the Gladstone Institutes of the University of California San Francisco. He received his Ph.D. from Carnegie Mellon University and was a postdoctoral fellow at Johns Hopkins University School of Medicine and Harvard University. He has edited several books, including three books for CRC Press.

Matthew R. Kaser is a Senior Partner at Bell & Associates in San Francisco and has been a part-time lecturer in the Department of Biological Sciences at California State University East Bay. He was on the faculty of the Department of Pediatrics, UCSF, an NIH Fellow at Habor-UCLA Medical Center, and held postdoctoral researcher positions at the University of California Irvine, M.D. Anderson Cancer Center in Houston, and Oxford University. With GCH, he co-authored a laboratory handbook on antibodies, also from CRC Press.

1 Puget Sound Then and Now

PHYSICAL DESCRIPTION

The Puget Sound is a complex fjord-estuary complex in Washington State and is connected to the Pacific Ocean by the Juan de Fuca Strait. In terms of its natural history, it is part of a much greater area that includes parts of British Columbia, such as Vancouver Island, and several large bodies of water, such as the Salish Sea, the Strait of Georgia, and Haro Strait. While we will focus primarily on the Sound itself, we will also take into account the geological history in a larger context.

The Puget Sound is about 160 km from north to south and stretches beyond the Canadian border in the north nearly to Oregon in the south (Figure 1.1). It is bounded on the west by the Olympic Peninsula and on the east by the Cascade Range. The average depth is about 140 m, and its deepest point is about 280 m. A sill at the Admiralty Inlet separates the Puget Sound from the Strait of Juan de Fuca and the Pacific Ocean. Sills are barriers on the sea floor that are relatively shallow but hinder the free circulation of water between the different bodies of water. The other connections to the Pacific are through Deception Pass and the Swinomish Channel, both of which empty into the Skagit Bay. The Sound itself is often divided into four basins: Whidbey Basin, Main Basin, South Sound, and Hood Canal. Sills separate the basins, except between the Whidbey and Main Basins. It has 3790 km of shoreline (Burke, nd).

The watershed covers nearly 43,000 km^2 and includes over 10,000 rivers and streams that drain into the Sound (Burke, nd). The major rivers that flow into the Sound include the Cedar/Lake Washington canal, Duwamish/Green, Elwha, Nisqually, Nooksack, Puyallup, Skagit, Skokomish, Snohomish, and Stillaguamish. Others flow down the western slope of the Olympic Mountains toward the Pacific.

The region is also a major metropolitan area with several large cities, such as Seattle, Tacoma, Renton, Everett, Bremerton, Olympia, and Federal Way. The larger area also includes the cities of Victoria and Vancouver in British Columbia. And because it is on the border between the United States and Canada, the Sound is a significant center of international trade.

NATURAL HISTORY

The natural history of the region involves tectonic forces, volcanism, glaciers, the rise and fall of sea levels, and the activities of humans over many millions of

DOI: 10.1201/9780429487439-1

FIGURE 1.1   Puget Sound region. The region is bounded by the Olympic Peninsula to the west, the Cascade Mountains to the east, British Columbia and Vancouver Island to the north and lowlands to the south. A number of significant rivers flow into the Sound, and the Sound itself is connected to the Pacific Ocean through the Strait of Juan de Fuca (Map courtesy of US Geological Survey).

years. This figure will be useful throughout the book). Plate tectonics are deeply involved in the past and future of the Puget Sound region. Their actions have been and will be altered by global climate change that brought changes in sea levels and massive glaciers that carved the Earth's surface.

For much of its history, most of Washington was under water. The surface of the Earth is divided into plates that "float" on the hotter layer under the crust, the mantle. New crust is created at mid-ocean ridges, and it pushes the plates around the planet. This process was deeply involved in the development of modern-day Washington state. It began with the collision of two of the Earth's massive tectonic plates. Over time, the heavier oceanic Farallon plate was subducted by the lighter continental North American plate. The Farallon plate originally covered the eastern half of the Pacific Ocean. Today, it is essentially completely subducted under the North American plate. However, a few small pieces of that plate continue to be subducted. Most important for the Puget Sound region is the Juan de Fuca plate. Part of that plate remains offshore of Washington, British Columbia, and Oregon and provides the energy that powers the Cascade volcanoes. As the Juan de Fuca plate slid under, large amounts of rocky material were scraped up off the bottom of the ocean and added to the edge of the North American plate. This heterogeneous mass included sandstone, shales, conglomerates, chert, basalt, and limestone.

While water is not quite so powerful a force as the tectonic plate movement, it has nevertheless had a significant role in sculpting the land of the Sound region. Changes in the global climate yielded lower and then higher ocean levels that alternately emptied and flooded the Sound, respectively. The changes from wet to dry to wet environments repeatedly over the last two million years caused changes in the land, plants and animals in the region. In addition, atmospheric rivers periodically drenched the region for millennia with fresh water that caused floods and erosion.

The most recent Ice Age covered the Puget Sound with a thick layer of ice just 17,000 years ago. The Cordilleran Ice Sheet covered the northern third of Washington state and essentially all of British Columbia. At present-day Seattle, the ice was 1000 m thick, and as it advanced and retreated seven times, it carved the Puget Sound.

Like all of the west coast of the United States and Canada, the Puget Sound is part of the Pacific Ring of Fire. As a result, the region experiences periodic earthquakes, some quite large. The largest earthquakes come from breaks alone the entire subduction boundary. The Ring of Fire also features volcanoes, and the Cascades include many volcanoes (USGS, nd). They affected the region in the past, and importantly, they are still active and will influence the region in the future. In fact, several of them have erupted within recorded history. Mount St. Helens erupted in 1980. Two of those volcanoes are located in the Puget Sound region. The most obvious is the beautiful Mount Rainer, which dominates the cities of Seattle and Tacoma (Figure 1.2). It last erupted about 1000 years ago, but it remains active and a threat to the region. Its previous eruptions have unleashed large lahars as well as explosive eruptions of tephra. Less well known

FIGURE 1.2 Mt. Rainier looms over Tacoma. While the mountain is amazingly beautiful, it is also quite dangerous. Although it has not erupted in 500 years, it is an active volcano, and its major threat is the lahars that would likely accompany an eruption (photograph courtesy of Lyn Topinka and US Geological Survey).

perhaps is the Glacier Peak just east of the Sound. Since the last Ice Age, it has featured the largest and most explosive eruptions in the continental United States, and it has erupted six times in that period.

Over millions of years, those plants and animals have evolved. Because much of the area was underway early on, only one dinosaur bone has been found. An 80-million-year-old thigh bone of a therapod was found in 2015. However, about ten million years ago, the Olympic Mountains began to be uplifted, but much of Washington and even the peaks of the Cascades were under ice. A large number of mammal bones have been found from the Quaternary, including the Columbia mammoth, giant ground sloth, a rhinoceros relative called *Diceratherium*, bison, mastodons, and a whale relative called *Chonecetus*. In addition, many trilobites and ammonites have been found. Most of those large mammals are also long extinct (Figure 1.3).

The Puget Sound has a diverse ecology and has been home to a wide variety of plants and animals since its formation. For example, it is a major stop on the Pacific flyway for migrating and resident birds. Port Susan and Skagit Bay host more than 50,000 birds each winter. Snow geese, Dunlin sandpipers, yellowlegs, dowitchers, and many more feast on the amazing biomass in the mud and sand of the Sound.

		Holocene	*0.012
Cenozoic	Quaternary	Pleistocene	2.58
	Neogene	Pilocene	5.33
		Miocene	23.0
	Paleogene	Oligocene	33.9
		Eocene	56.0
		Paleocene	66.0
Mesozoic	Cretaceous	Upper	100.5
		Lower	145.0

FIGURE 1.3 Geologic timeline. The various geologic eras will be referred to throughout much of the book.

Native Americans arrived in the region at least 10,000 years ago. They made their way from Asia during the last Ice Age when sea levels were lower than today, and the coast was 15–50 km west of where it is today. When that coldest part of the Ice Age ended, sea levels began to rise. Year after year, the tides were higher than those before. The ocean was creeping further and further onto the land. Those early Native Americans likely saw the ice cover of the last Ice Age recede and the water return to parts of the Puget Sound.

Sadly, natural systems have too often been the losers since humans arrived in the Sound region. In the last couple of hundred years, human activity has greatly influenced the Sound region, and much of that influence has not been for the better. Small towns have grown to become major cities, and the population continues to increase. More people means more stress on the natural environment to provide housing, transportation, and other systems in the region. Bridges, roads, ports, water and sewage systems and structures were built. Wetlands were filled and developed. Rivers were channeled and the shore was armored. In addition, human activities have introduced quite a number of invasive species that have disrupted natural populations and, in some cases, led to their extinction.

There is no magic wand to wave to restore the region to what it once was, but it is not all bad news. In more recent decades, people have come to realize that the natural habitat is important for people too. For too many years, the wetlands and the Sound have been treated like dumps for whatever waste the residents wanted to dispose of, or they were seen as ripe for development. Now efforts are underway to restore wetlands and other wild areas for the benefit of plants and animals, and also for humans.

Change is inevitable. In the crush of our daily activities, it's easy to forget that the forces that built the Sound are not influenced by our meetings, jobs, smartphones, or desires. They move at their own speed and come in their own time. The Sound was not always here, and it will not always be here in the future. The forces that created it are still active and will one day destroy it. Even now, we are feeling the effects of the climate crisis with much hotter temperatures and seemingly endless fire seasons. Sea levels are rising as the Earth's large icecaps and glaciers melt. That extra water will change the coastline and region over the next years.

We will describe the natural history and evolution of the Puget Sound region (e.g., its geology, plants and animals, people and their activities, and the connections among these) over the last 50 million years through the present and into the future. We will explain how those forces that built the Bay are still active today. Scientists have already made predictions of what might happen in the future. The most powerful are the geologic forces, and the region has been subjected to considerable tectonic activity over the years. Several faults crisscross the region and surrounding area, resulting in periodic earthquakes and other ground movements. Other forces include the rise and fall of the ocean with changes in climate, atmospheric rivers, and finally humans. Another Ice Age is not out of the question, but less likely. We will also review the predictions about the future of the Bay region and how those forces will eventually change the region beyond recognition. Humans witnessed the end of the Ice Age, the rise of sea levels, and the flooding of the Sound. We do not know if humans will see the end of the Sound, but we do know that change is coming.

REFERENCES

Burke (nd) *Welcome to the Puget Sound*. The Burke Museum. Retrieved from: https://www.burkemuseum.org/static/FishKey/aboutps.html#:~:text=Over%2080%25%20of%20surface%20water,about%20Puget%20Sound%2C%20click%20here%20. March 20, 2021.

USGS (nd) *Cascade Volcano Observatory*. US Geologic Service. Retrieved from: https://www.usgs.gov/observatories/cascades-volcano-observatory. March 20, 2021.

2 Geological Origins of the Puget Sound

BUILDING WESTERN WASHINGTON

The Puget Sound is a relatively young estuary system. About 200 million years ago, the West Coast of North America was just west of Pullman and Spokane today. The western part of the state had its beginnings deep in the Pacific Ocean, and the forces that caused it to form are related to how the continents have formed and evolved over millions of years. Those forces include plate tectonics, volcanic activity, crustal uplift and depression, folding, and erosion. The result is a complex jumble. Interestingly, those forces are still active today.

To all of us, the ground we stand on seems solid. In fact, it isn't. It is in constant motion and ever changing. The energy comes from the Earth's superhot core. The inner core is a solid mass of mostly iron at 4200–7200°C. The outer core is molten iron and nickel at 3200–5200°C. Between the core and the crust is the mantle, which is also extremely hot and continually moving. The crust makes up only a relatively thin layer on top of the mantle (about 30 km thick on the continents and about 5 km under the oceans). It is divided into massive plates that "float" on the mantle far below the surface. The movement of the plates is powered by the convection in the mantle. This forces new material up through weak areas in the crust, such as the spreading areas or rifts in the middle of the oceans. New material upwells from the mantle, is pushed along over the surface, and is later subducted where plates meet. Along those edges, earthquakes and volcanoes are common. Ultimately, the energy for the movements comes from the Earth's cooling, but still, fantastically hot mantle and core.

PLATE MOVEMENT

The crust is divided into plates of varying sizes. Some are oceanic, and others also carry the continents. Movements of the tectonic plates are caused by movements in the mantel due to the heat there. The interactions at the boundaries of the plates are of three types: convergent, divergent, and transform. The convergence can be oceanic-oceanic, continental-continental, or oceanic-continental. In oceanic-oceanic convergences, one or both of the plates will override the other and be lost by subduction. In continental-continental, they run directly into each other to produce a crumpled boundary that creates large mountain ranges, such as the Himalayas. In oceanic-continental, the denser oceanic is subducted under the lighter continental crust. Divergent plates move away from each other. In transform boundaries, the two plates slip along against

DOI: 10.1201/9780429487439-2

each other. The creation of California first involved subduction and later a transform boundary.

The continental plates have alternately existed as a huge single supercontinent or multiple smaller continents. The supercontinents generally last about 100 million years before they break up and the resulting continents drift apart. Every few hundred million years, the cycle repeats itself. At least three supercontinents have existed over the past two billion years. There is some evidence for others, but it is not compelling. Nuna is the oldest known supercontinent known so far. It was in place about 1.8 billion years ago. The next was Rodinia about one billion years ago, and finally, Pangaea was about 300 million years ago. At the other times between those three, smaller continents, such as we have today, existed. Scientists use the magnetization of iron to show how those continents interacted in a supercontinent. In molten rock, magnetized iron oriented itself along the magnetic lines of the Earth. Once the rock solidified, its history was locked in stone and can be compared to the magnetic orientation of other rocks. By comparing iron-containing rocks from different regions and knowing the time frame, scientists can put the puzzle back together again and determine the movement of the continents and their joining to form a supercontinent.

About 300 million years ago, the supercontinent of Pangea broke up to form the Atlantic Ocean, and the North American plate began migrating west. At the same time, the heavier oceanic Farallon plate was moving east. At their intersection, the Farallon plate was subducted under the North American plate.

SUBDUCTION

As the new material is forced to the surface at mid-ocean rift zones, it pushes the plates away from the rift in different directions. The new rock begins as pillow basalt that erupted thousands of kilometers out in the Pacific Ocean at the mid-ocean rift 100–200 million years ago. That material entered a "conveyor belt" that moved it from its origin to the coast over millions of years.

That conveyor belt pushed the massive Farallon plate eastward, where it collided with the North American plate. As oceanic plates are denser than continental plates, the Farallon plate sank below North American plate in a process called subduction. Subduction also creates enormous amounts of heat as the plates grind past one another deep underground. That heat can yield volcanic activity at some distance inland from the plate boundaries.

Also at the boundary, large pieces of rock called terranes broke off from the subducting plate. Terranes are pieces of crust that is lower in density. As its parent plate is subducted, it becomes detached and is accreted onto the other plate. In this way, crustal material is transferred directly from one plate to another. The transferred material might also have been old island arcs.

In the Cenozoic era, the Crescent Terrane was added to the North American coast. It includes much of the northern and eastern parts of the Olympic Peninsula. Even today, more of the ocean floor is being pushed into the Coast Ranges at the leading edge of the subduction zone.

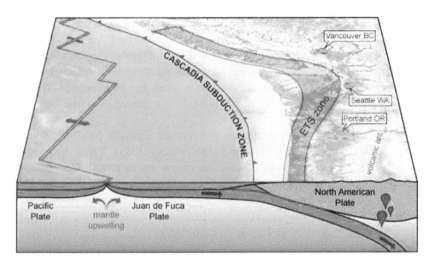

FIGURE 2.1   Cascadian subduction zone. The basic structure of the region is the result of the subduction or the Juan de Fuca plate underneath the far-larger North American plate (Diagram courtesy of the US Geological Survey).

As the Juan de Fuca plate, a small left-over piece of the Farallon plate, began to be subducted under the North American plate, the Cascade Episode began (Figure 2.1). The subduction process created a chain of volcanoes under the North American plate that have been erupting for 36 million years. About 12–17 million years ago, a huge mass of molten rock broke through cracks in the crust and resulted in the basalts of the Columbia Plateau. The Cascade Range rose up in the last five to seven million years. These mountains are a group of volcanoes, resulting from plate subduction. The subduction causes an uplift of the leading edge of the North American plate. Over time, that uplift is often balanced by erosion. Pazzaglia and Brandon (2001) examined terraces from the late Quaternary in the Clearwater River on the Olympic Peninsula. They found that the area had reached a steady state between uplift and erosion. Uplift also contributed significantly to the growth of the Range (Reiners et al., 2002). The uplifts of the Cascade and the Olympic Mountains resulted in a low area or trough between the mountains. Water filled that area to form Puget Sound.

Slab Rotation and Rollback

During subduction, multiple forces contribute stress to deform the surface of the overriding plate. Slab rollback or rotation is one of these (Cassel et al., 2018). In slab rollback, the subducting plate bends back to migrate in the opposite direction of the overall plate movement. These actions can cause the surface to be uplifted or depressed. Those forces include magmatism, heating, delamination, and mantle flow.

In the Cascade region, the Juan de Fuca plate continues to be subducted under the North American plate. At the same time, the Juan de Fuca plate is caught between the much larger North American and Pacific plates. The forces involved are causing large blocks of material to move together, as if they are locked (McCaffrey et al., 2007). Much of the stress is taken up by spreading of the Juan de Fuca Ridge and subduction, but about 20–25% is passed on to the overriding continental plate.

Interestingly, the Miocene Cascade volcanic arc seems to be slowly rotating in a clockwise direction (Wells and McCaffrey, 2013). From its position 16 million years ago, the northern section of the arc in British Columbia has migrated eastward, and the southern section in Oregon has moved westward. In the north, the block rotation and slab rollback are in opposite directions. In the southern section, back arc block rotation and slab rollback are in the same direction. The rotation is small (1.0°/million years).

EARTHQUAKES AND MAJOR FAULTS

The Puget Sound region is extremely seismically active, and three types of earthquakes are possible. First, the greatest threat to the region is a subduction earthquake, resulting from slippage along the entire Cascadia subduction zone. Such an earthquake occurred in 1700 and reached a magnitude of 9.0. Although a subduction earthquake releases a great deal of energy, it is spread across a wide area. Second, intraslab earthquakes in the Benioff zone are caused by slippage on a small part of the subducting plate. They occur at a depth of about 50 km and can result in an earthquake of magnitude 6–7. Finally, crustal earthquakes involve the many faults in the region. These occur at depths of about 25 km and have magnitudes of 6–7. They are not nearly as strong as the subduction zone earthquakes, but they can do a lot of damage because they are close to the surface.

One of the most active areas is under the Puget Sound between Olympia and the Southern Whidbey Island Fault (Stanley et al., 1999). This area coincides with the southernmost extent of the glaciation, and the high degree of seismicity might be related to the continuing rebound of the land from the weight of the glacial ice.

The easiest way to identify a fault is by a scarp, the scar remaining after a surface rupture caused by a break along the fault. The Puget Sound has a multitude of faults since it sits on the part of the North American plate that is overriding the remnants of the subducting Juan de Fuca plate. However, the faults in the region were long obscured by layers of sediment left by the glaciers that covered much of Washington thousands of years ago. In addition, significant development of the area added another layer to camouflage the faults. Fortunately, more recent airborne laser surveys, such as LIDAR or laser detection and ranging, have revealed more detail of the faults in that area.

SEATTLE FAULT

The Seattle fault runs under the city. Surprisingly, its potential was only recognized in 1992 when evidence of a significant earthquake from 1100 years ago was found. This fault and the Tacoma fault are the most dangerous in the region. Estimates show that the fault can generate an earthquake of magnitude 7.0 that would cause serious damage to many of the masonry structures in the city. Furthermore, it would likely cause a tsunami of 2 m on Elliott Bay that would inundate many low-lying areas. The fault runs for 70 km from Fall City to the South Whidbey Island fault near Hood Canal and is 4–7 km wide. Its origin is related to the Seattle Uplift, which pushed up when a large block of basalt from the Crescent Formation 50–60 million years ago.

High-resolution imaging has been used to examine Holocene lake deposits in Lake Washington (Karlin et al., 2004). These identified a number of submarine landslides that show strands of the Seattle fault system. The evidence showed more than one significant event in the last 11,000 years. A large earthquake in 900 AD caused a 7-m uplift (Ludwin et al., 2005). Radiocarbon dating and magnetic profiles show that the sediments were disturbed seven times in the last 3500 years, suggesting that a large earthquake has occurred in the Puget Sound region every 300–500 years. Additional evidence for a large earthquake about 1000 years ago comes from a study of trees submerged in Lake Washington. By correlating tree ring data with other data, Jacoby et al. (1992) showed that an earthquake and tsunami occurred at that time. Arcos (2012) examined the sediments at the end of Sinclair Inlet and found at least 3 m of uplift occurred before a tsunami there. The findings indicate that the earthquake was most likely on the Seattle fault than the Tacoma fault. The tsunami traveled 2 km down the Gorst Creek valley and left a 40-cm-thick deposit there.

Interestingly, the legends of Native Americans in the area speak of a dangerous serpent-like creature that lives in the ground called a'yahos. Ludwin et al. (2005) found that these legends are associated with ancient earthquakes in the Seattle area. Some seem to relate specifically to the 900 AD event. Thus, the science may corroborate some legends, and the legends may provide clues to the researchers about the geologic history of this and other areas.

TACOMA FAULT

This fault lies in the southern part of the Puget Sound and traverses populated areas just north of Tacoma for about 50 km. It could cause a magnitude 7.1 earthquake (Gomberg et al., 2010), a tsunami, and liquefaction that would result in serious damage to the region. The entire region is undergoing transpression that causes uplifts, basins, and reverse faults that cross the Sound (Sherrod et al., 2004). The Tacoma fault is the southern boundary of an area of uplift (i.e., the Seattle uplift). The northern boundary is the Seattle fault. Calculations show that Tacoma would be damaged more by a tsunami resulting from an earthquake on the Seattle than the Tacoma fault (Walsh et al., 2009). This is because one from the Seattle fault would pass over deeper water and thus would yield much greater amounts of water.

SOUTHERN WHIDBEY ISLAND FAULT

The fault marks a boundary between pre-Tertiary crystalline rocks and Tertiary coast range basalts and is covered by glacial sediments that obscure its track (Sherrod et al., 2008). The Rattlesnake Mountain fault is now considered to be a southern part of the Southern Whidbey Island fault, making the entire length of the fault system about 150 km. In fact, it might extend beyond that point to join with or cross the Olympic-Wallowa Lineament. There is evidence of earthquakes of magnitude 6.5–7.0 from 3000 years ago. Others found four to nine events on the fault in the last 16,400 years (Sherrod et al., 2005).

OTHER FAULTS

Several other faults are also located in the Puget Sound region. These include the Rogers Belt, Cherry Creek, Saddle Mountain, Doty, Saint Helens, Devils Mountain, and Strawberry Point faults.

VOLCANOES

There are lots of volcanoes in the Cascades. O'Hara et al. (2020) counted 2835 volcanoes that have erupted at least once in the last 2.6 million years. Of those, 231 are still active and erupted in the last 10,000 years.

Five large active volcanoes are located in Washington State (i.e., Baker, Rainier, Mount St. Helens, Adams, and Glacier Peak). At least, three of them (i.e., Rainier, Baker, and Glacier Peak) can be considered to be part of the Puget Sound Region. All are part of the Cascade Mountains. The energy for the volcanoes comes from the active subduction of the remains of the Juan de Fuka plate under the North American plate. The friction of the two plates grinding past one another creates enormous amounts of heat that melts the subsurface rocks until they melt and rise up to the surface.

The volcanos are snow-capped and beautiful, but they are also very active volcanoes and dangerous. The most recent eruption of Mount St. Helens in 1980 provided an unusual demonstration of how many of the Cascade volcanoes have erupted over the last millions of years to build the Puget Sound region. The mountain began building about 13,600 years ago, and it has gone through several periods of activity and dormancy. That story is similar for many of the Cascade volcanoes at some point in their lifespan. The recent catastrophic eruption featured a massive landslide and a lateral or side-ways explosion that blew away much of the mountain. About 400 m of the summit (about 2.5 km^3 of material) was lost in the explosion. A crater about 1.6 km wide was left (Figure 2.2).

FIGURE 2.2   Mount St. Helens lahar sediments (Photograph courtesy of Kurt Spicer, US Geological Survey).

DANGERS FROM VOLCANOES

Volcanoes pose a number of dangers. The dangers include poison gases, lava, pyroclastic flows, lahars, and explosions. Many Cascade volcanoes release gas continuously (USGS, nd-a). They include water vapor, carbon dioxide, sulfur dioxide, and hydrogen sulfide. Some are poisonous to animals and plants. However, the gases released from a volcano contain information about the state of the volcano (USGS, nd-b). Carbon dioxide comes from magma deeper in the volcano, and so, a rise in the levels of that gas indicates fresh magma moving up. Hydrogen sulfide typically comes from quiet volcanoes, but sulfur dioxide indicates a more active phase for the volcano.

Lava is molten rock or magma that is extruded from a volcano. The temperature of the lava is 800–1200°C. The lava can form fountains or it can flow out. It can be runny or viscous. Lava destroys buildings and other infrastructure, but it moves so slowly that it is rarely a threat to people. In the Cascade volcanoes, the lava rarely spreads beyond the actually volcano, but lava bombs can be hurled for considerable distances. Lava fields can be extremely dangerous. The cooling lava can look solid, but it also might be a fragile thin covering over still-hot lava.

Pyroclastic flows have been common during eruptions of the Cascade volcanoes (Gardner et al., 1998). They result from specific types of volcanic eruptions and are very dangerous. They consist of dense, fast-moving flows of solidified lava pieces, volcanic ash, and hot gases. The mixture is extremely hot. The temperatures reach 1000°C. They are essentially impossible to outrun since their speeds are as high as 200 m/s. They can rapidly change the landscape. They burn all living material and leave a layer of solidified lava and thick volcanic ash. They block streams, reroute rivers, cause flooding, and set off lahars. Pyroclastic flows buried the cities of Pompeii and Herculaneum in 79 AD. The people there were instantaneously fossilized by the burning material.

Lahar is an Indonesian word for the mudflows that accompany volcanic eruptions if sufficient water is available to be rapidly mobilized. Lahars are the greatest threat from the volcanoes in the Puget Sound region (USGS, nd-c). They carry masses of rock, mud, and water rapidly downstream. The Cascade volcanoes are particularly susceptible to lahars (Figure 2.3). They have steep sides and plenty of water stored as ice in their glaciers. Lahars are very dangerous in that they move very fast and carry very large amounts of debris down river valleys. They can also travel great distances from the actual eruption. The debris flow has the consistency of motor oil or wet concrete. The speed of the flow can be more than 20 m/s and 10 m/s for more

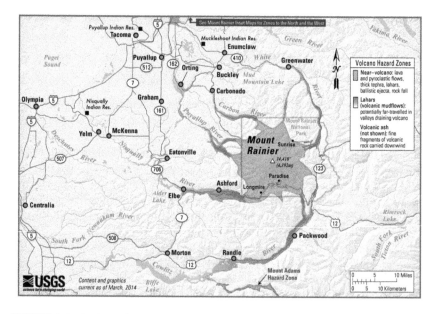

FIGURE 2.3 Lahars from Mount Rainier. The greatest danger of an eruption by Mount Rainier is lahars. They flow downhill at high speed, and the mixture of dirt, rock, and water has the consistency of wet cement. The map shows the likely path of the lahars (Map courtesy of US Geological Survey).

than 50 km from the source. The force can move even multi-ton objects, such as boulders, logs, and cars. Once the flow ends, the material can encase objects in something like concrete. Lahars change the environment dramatically.

MOUNT RAINIER

At 4392 m, Mount Rainier is a beautiful and dangerous active volcano that dominates the view of Seattle and Tacoma. It is the tallest mountain in Washington and in the Cascade Range. It is heavily glaciated and provides the source of five major rivers. Mount Rainier is a stratovolcano because if features steep sides and a summit crater. Lava from stratovolcanoes typically has a high viscosity and so does not flow far before solidifying. Stratovolcanoes are sometimes called composite volcanoes because the cooled lava builds up layers on the cone. The summit actually features two partially overlapping craters, each 300 m in diameter.

The mountain has 26 major glaciers that are the source of many of the region's rivers. The Carbon, Puyallup, Mowich, Nisqually, White, and Cowlitz Rivers all begin at glaciers on the mountain. Several of those rivers flow to the Puget Sound, Commencement Bay, and the Columbia River. The paths of these rivers are important because lahars follow river valleys. The thickness and volume of glacial ice have been reduced in recent years (Sisson et al., 2011). All the glaciers have thinned except for the Emmons and Winthrop glaciers. About 14% of all the ice was lost from 1970 to 2008.

The mountain has not erupted for a while, but there is plenty of seismic activity there (Shelly et al., 2013). Two to three earthquakes occur every month, and occasionally, swarms of small earthquakes are recorded. Swarms were seen in 2002, 2004, and 2007. More than 1000 were detected over three days in September 2009. Their significance is unknown. They might be due to slippage, increases in fluid pressure, or intrusion of magma.

The volcano began about 840,000 years ago. McGary et al. (2014) described how the volcano might have grown. They surmise that the subducting slab heats and melts near the mantle, and the material rises to penetrate the crust. It grew by adding layers of lava, debris, and pyroclastic flows. The cone of today is 500,000 years old. Over that time, the mountain has been alternately growing with eruptions and shrinking with erosion. The mountain has erupted 10–12 times in the last 2600 years (Sisson and Vallance, 2009). Most of the eruptions were weakly explosive, but at least two had block-and-ash pyroclastic flows. The most recent eruption with lava was 1000 years ago.

The collapse of sections of the volcano generally occurs on flanks that have been weakened by hydrothermally altered rock (Watters et al., 2000; Reid et al., 2001). The upper west slope of Mount Rainier is most susceptible to these collapses. Part of the height has been rebuilt by subsequent eruptions. An avalanche 5000 years ago reduced the height of the mountain by about 500 m and caused the Osceola Mudflow (Vallance and Scott, 1997).

The mudflow was preceded by several eruptions with pyroclastic flows (Vallance and Scott, 1997). Then a quiet period lasted for hours to hundreds of years before the big event. The avalanche was accompanied by a huge lateral eruption and explosion. The mudflow was massive and created a path of destruction all the way to the current-day city of Tacoma. The Osceola Mudflow was a very large lahar that flowed from Mount Rainier down the west and main forks of the White River to the Puget Sound near Auburn. The mudflow moved about 2–3 km^3 of material downstream and buried much of the Puget Sound lowland. The amount of material is hard to imagine. The height of the flow was reduced rapidly as it moved away from the volcano. Near the volcano, the levels were 400 m above the valley floor, but it was still on average 100 m about the floor 5 km from the volcano. Much of the present King and Pierce Counties is built on those deposits. Other mudflows have taken place. For example, about 550 years ago, another smaller mudflow, the Electron Mudflow, occurred. Debris flows do not depend only on eruptions. They can occur from climate change, retreating glaciers, and storm patterns (Legg et al., 2014). So the debris flows are an ongoing hazard around Mount Rainier.

Mount Rainier is a very dangerous volcano. An eruption is always a threat, especially one accompanied by a significant lahar. Climate change is also thinning the glaciers on the mountain, and that increases the risk for a lahar (Tuffen, 2010). A large number of people now live in the traditional paths of those lahars.

Mount Baker

Mount Baker is a stratovolcano about 48 km due east of Bellingham, Washington. It is the youngest volcano in the Cascades. It is likely only 140,000 years old and possibly no older than 80,000–90,000 years. The current cone of 3286 m is located on top of older cones that were mostly eroded away in the last ice age. Like most Cascade volcanoes, Mount Baker is heavily glaciated, and it is losing those glaciers. From 1984 to 2009, the glacier retreated 240–520 m (Pelto and Brown, 2012). Unlike most other Cascade volcanoes, Mount Baker does not typically feature explosive eruptions. Over its existence, the volcano has erupted at least 70 times (Hildreth et al., 2003). Many of the features of the volcano's history can be easily seen in the Baker River Valley (Tucker et al., 2007).

Glacier Peak

Glacier Peak is an active stratovolcano with an extensive history of eruptions. It has erupted five times in the last 3000 years. The last time was 1100 years ago. Its height is 3213 m (USGS, nd). The age of the volcano is poorly understood. Extensive glaciation and subsequent eruptions have eroded and buried the earliest cones. Estimates are that it is 200,000–600,000 years old. Glacier Peak and Mount St. Helens featured large explosive eruptions in the past 15,000 years. During these eruptions from Glacier Peak, massive amounts of volcanic ash were blown eastward. Glacier Peak has extensive tephra beds that represent a number of Plinian and sub-Plinian eruptions (Plinian eruptions involve great clouds of superheated ash). After studying these beds, Kuehn et al. (2009) assigned a date of 11,600 by ^{14}C dating and a calibrated age of 13,710–13,410 years ago. The mountain also has a dozen glaciers, and its eruptions have included lahars in the surrounding river valleys.

VISIBLE REMINDERS OF THE FORCES THAT BUILT THE PUGET SOUND

The Puget Sound region has a rich geologic history, and the remains of many of those processes can be seen all around the Sound (Matthews, nd).

MOUNT RAINIER

This large active stratovolcano dominates the view from Tacoma and Seattle. It is 4392 m tall and the highest mountain in the Cascades. It is also as dangerous as it is beautiful (Reid et al., 2001). The summit of the volcano is covered by a glacier. It has 26 major glaciers, and the ice covers 93 km^2. The ice in the glaciers has great potential to produce huge lahars that threaten the entire Puyallup River valley, where 80,000 people live. The mountain consists of old lava flows, debris flows, and pyroclastic ejecta. It began about 840,000 years ago, is made of mostly andesite, and is heavily eroded today. About 5000 years ago, the peak collapsed during an eruption. That eruption also produced the Osceola mudflow that carried massive amounts of debris nearly to Tacoma and the Sound (Figure 2.4).

PILLOW BASALTS IN THE OLYMPIC NATIONAL FOREST

The rocks erupted under the Pacific Ocean about 55 million years ago. These may be a terrane that was accreted onto the North American plate from the subducting oceanic plate. They might have resulted from a slab window. Alternatively, they might have been caused by lava spilling out onto the floor of the North American plate. The volume of the lava might be as much as 100,000 km^3.

ERRATICS

In geology, erratics are large objects that have been transported from one place to another. Glaciers often move very large rocks, and several erratics can be seen on the beach at Seattle's Discovery Park. Other erratics can be found around the area, including the "Big Rock" at Duvall and another one in the Wedgwood neighborhood. A large erratic in Leschi Park in Seattle features many visible fossils. A granodiorite erratic is located in Seattle's Ravenna Park. Granodiorite erratics probably originated in British Columbia and were carried along by the Cordilleran ice sheet.

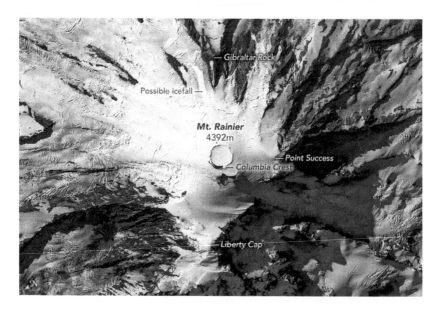

FIGURE 2.4   The glaciers of Mount Rainier. The volcano is covered with glaciers. The mountain has been growing for about half a million years. The energy for the volcano comes from the subduction of the Juan de Fuca plate beneath the North American plate. It has erupted about a dozen times over the last 2600 years (Photograph courtesy of the National Aeronautics and Space Administration).

Sea Stacks

The coastline of Olympic National Park is quite rugged. The sea stacks are one of its most beautiful and mysterious features. Millions of years ago, volcanic intrusions penetrated the layers of sedimentary rocks. Over time, the softer mudstone eroded leaving the stacks of harder volcanic rock standing (Figure 2.5).

Alpine Lakes

The movement of glacial ice carved these depressions, and the melting ice filled them with water.

Whidbey Pleistocene Stratigraphy

Blowers Bluff on Whidbey Island shows sediments from the interglacial times that are overlain by glacial till and glaciomarine drift. The oldest Pleistocene sequence in this part of the Puget lowland is found at Useless Bay and Double Bluff. The sediments are deformed into swirls and whorls and penetrated by clastic dikes.

FIGURE 2.5   Sea stacks on Third Beach (Photograph courtesy of the National Park Service).

MAZAMA ASH DEPOSIT

In the Skagit River Valley between Marblemount and Newhalem, about 10 m of volcanic ash that erupted from Mount Mazama in the Crater Lake caldera collapse is clearly visible. The ash is nearly pure rhyolite and thus is shards of siliceous glass.

SERPENTINE OUTCROPS

In Washington Park near Anacortes, outcrops of serpentine can be seen. Serpentine is a metamorphic rock that likely originated in the Earth's mantle. It is green and soft and has a sheen to it. The deposits here were squeezed out from other rocks.

DUNGENESS SPIT

At the north end of the Olympic Peninsula just north of Sequim, Dungeness Spit is a long slender spit of natural sand that sticks out into the Strait of Juan de Fuca. It was originally formed when the glaciers retreated. It also gives a spectacular view of the Strait that connects the Puget Sound to the Pacific Ocean.

Mima Mounds

The Mima Mounds are near Olympia and contain thousands of rounded mounds about 2 m high and 5 m wide. Originally, they covered over 30 km. Their cause is completely unknown, but they also include a stand of native prairie.

Lahars

Eruptions on Mount Rainier have melted large amounts of the glacial ice on the mountain. The melt water then careens down the mountain carrying massive amounts of debris with it. The damage can be catastrophic. About 5600 years ago, a huge lahar carried debris all the way to the Puget Sound. The cities of Puyallup, Buckley, Auburn, and Sumner are built on material transported by the Osceola Mudflow.

FORCES BUILDING THE SOUND

All of these geological forces working over millions of years resulted in the Puget Sound that we know today. Geology with its plate movements, volcanoes, and earthquakes is the greatest force involving in the making and unmaking of the Sound. However, other forces, such as water (Chapter 3) and humans (Chapter 6) also contributed.

REFERENCES

Arcos MEM (2012) The A.D. 900–930 Seattle-fault-zone earthquake with a wider co-seismic rupture patch and postseismic submergence: Inferences from new sedimentary evidence. *Bulletin of the Seismological Society of America* 102: 1079–1098.

Booth DB, Haugerud RA, Troost KG (2003) The geology of Puget Lowland rivers. In: Montgomery DR, Bolton S, Booth DB, Wall L (editors). *Restoration of Puget Sound Rivers*, pp. 14–45. University of Washington Press, Seattle, WA.

Cassel EJ, Smith ME, Jicha BR (2018) The impact of slab rollback on Earth's surface: Uplift and extension in the hinterland of the North American Cordillera. *Geophysical Research Letters* 45: 10,996–11,004.

Gardner JE, Carey S, Sigurdsson H (1998) Plinian eruptions at Glacier Peak and Newberry volcanoes, United States: Implications for volcanic hazards in the Cascade Range. *GSA Bulletin* 110: 173–187.

Gomberg J, Sherrod B, Weaver C, Frankel A (2010) *A Magnitude 7.1 Earthquake in the Tacoma Fault Zone: A Plausible Scenario for the Southern Puget Sound Region, Washington*. Geological Survey Fact Sheet, Washington. p. 4.

Hildreth W, Fierstein J, Lanphere M (2003) Eruptive history and geochronology of the Mount Baker volcanic field, Washington. *GSA Bulletin* 115: 729–764.

Jacoby GC, Williams PL, Buckley BM (1992) Tree ring correlation between prehistoric landslides and abrupt tectonic events in Seattle, Washington. *Science* 258: 1621–1623.

Karlin RE, Holmes M, Abella SEB, Sylwester R (2004) Holocene landslides and a 3500-year record of Pacific Northwest earthquakes from sediments in Lake Washington. *GSA Bulletin* 116: 94–108.

Kuehn SC, Froese DG, Carrara PE, Foit Jr FF, Pearce NJG, Rotheisler P (2009) Major- and trace-element characterization, expanded distribution, and a new chronology for the latest Pleistocene Glacier Peak tephras in western North America. *Quaternary Research* 71: 201–216.

Legg NT, Meigs AJ, Grant GE, Kennard P (2014) Debris flow initiation in proglacial gullies on Mount Rainier, Washington. *Geomorphology* 226: 249–260.

Ludwin RS, Thrush CP, James K, Buerge D, Jonientz Trisler C, Rasmussen J, Troost K, de los Angeles A (2005) Serpent spirit-power stories along the Seattle fault. *Seismological Research Letters* 76: 426–431.

Matthews D (nd) *Northwest Geology Field Trips*. Northwest Geology. Retrieved from: https://nwgeology.wordpress.com/field-trip-locations/. March 28, 2021.

McCaffrey R, Qamar AI, King RW, Wells R, Khazaradze G, Williams CA, Stevens CW, Vollick JJ, Zwick PC (2007) Fault locking, block rotation and crustal deformation in the Pacific Northwest. *Geophysical Journal International* 169: 1315–1340.

McGary RS, Evans RL, Wannamaker PE, Elsenbeck J, Rondenay S (2014) Pathway from subducting slab to surface for melt and fluids beneath Mount Rainier. *Nature* 511: 338–340.

O'Hara D, Karlstrom L, Ramsey DW (2020) Quaternary volcanism in the Cascades arc. *Geology* 48: 1088–1093.

Pazzaglia FJ, Brandon MT (2001) A fluvial record of long-term steady-state uplift and erosion across the Cascadia forearc high, Western Washington State. *American Journal of Science* 301: 385–431.

Pelto M, Brown C (2012) Mass balance loss of Mount Baker, Washington glaciers 1990–2010. *Hydrological Processes* 26: 2601–2607.

Reid M, Sisson TW, Brien DL (2001) Volcano collapse promoted by hydrothermal alteration and edifice shape, Mount Rainier, Washington. *Geology* 29: 779–782.

Reiners PW, Ehlers TA, Garver JI, Mitchell SG, Montgomery DR, Vance JA, Nicolescu D (2002) Late Miocene exhumation and uplift of the Washington Cascade Range. *Geology* 30: 767–770.

Shelly DR, Moran SC, Thelen WA (2013), Evidence for fluid-triggered slip in the 2009 Mount Rainier, Washington earthquake swarm. *Geophysical Research Letters* 40: 1506–1512.

Sherrod BL, Blakely RJ, Weaver CS, Kelsey HM, Barnett E, Liberty L, Meagher KL, Pape K (2008) Finding concealed active faults: Extending the southern Whidbey Island fault across the Puget Lowland, Washington. *Journal of Geophysical Research* 113: B05313.

Sherrod BL, Blakely RJ, Weaver CS, Kelsey H, Barnett E, Wells R (2005) *Holocene Fault Scarps and Shallow Magnetic Anomalies Along the Southern Whidbey Island Fault Zone Near Woodinville, Washington*. Geological Survey, Washington. pp.S51C–S51022.

Sherrod BL, Brocher TM, Weaver CS, Bucknam RC, Blakely RJ, Kelsey HM, Nelson AR, Haugerud R (2004) Holocene fault scarps near Tacoma, Washington, USA. *Geology* 32: 9–12.

Sisson TW, Robinson JE, Swinney DD (2011) Whole-edifice ice volume change AD 1970 to 2007/2008 at Mount Rainier, Washington, based on LiDAR surveying. *Geology* 39: 639–642.

Sisson T, Vallance J (2009) Frequent eruptions of Mount Rainier over the last ~2,600 years. *Bulletin of Volcanology* 71: 595–618.

Stanley D, Villaseñor A, Benz H (1999) *Subduction Zone and Crustal Dynamics of Western Washington: A Tectonic Model for Earthquake Hazards Evaluation*. US Geological Survey, Washington. Open-File Report 99–311.

Tucker DS, Scott KM, Lewis DR (2007) Field guide to Mount Baker volcanic deposits in the Baker River valley: Nineteenth century lahars, tephras, debris avalanches, and early Holocene subaqueous lava. *The Geological Society of America Field Guide* 9.

Tuffen H (2010) How will melting of ice affect volcanic hazards in the twenty-first century? *Philosophical Transactions 1331 of the Royal Society A: Mathematical, Physical and Engineering Sciences* 368: 2535–2558.

USGS (nd) *Glacier Peak.* US Geologic Survey. Retrieved from: https://www.usgs.gov/volcanoes/glacier-peak. April 6, 2021.

USGS (nd-a) *Volcanic Gas Monitoring Gives Clues about Magma Below. Cascade Volcano Observatory.* US Geological Survey. Retrieved from: https://www.usgs.gov/observatories/cascades-volcano-observatory/volcanic-gas-monitoring-gives-clues-about-magma-below. March 29, 2021.

USGS (nd-b) *Volcanic Gas Monitoring at Mount St. Helens. Cascade Volcano Observatory.* US Geological Survey. Retrieved from: https://www.usgs.gov/volcanoes/mount-st-helens/volcanic-gas-monitoring-mount-st-helens?qt-science_support_page_related_con=2#qt-science_support_page_related_con. March 29, 2021.

USGS (nd-c) *Lahars – The Most Threatening Volcanic Hazard in the Cascades. Cascade Volcano Observatory.* US Geological Survey. Retrieved from: https://www.usgs.gov/observatories/cascades-volcano-observatory/lahars-most-threatening-volcanic-hazard-cascades#:~:text=Lahars%20transform%20the%20landscapes%20around%20Cascade%20Volcanoes.&text=Small%20debris%20flows%20are%20common,a%20few%20miles%20down%20valleys. March 29, 2021.

Vallance JW, Scott KM (1997) The Osceola Mudflow from Mount Rainier: Sedimentology and hazard implications of a huge clay-rich debris flow. *GSA Bulletin* 109: 143–163.

Walsh TJ, Arcas D, Venturato AJ, Titov VV, Mofjeld HO, Chamberlin CC, Gonzále FI (2009) *Tsunami Hazard Map of Tacoma, Washington: Model Results for Seattle Fault and Tacoma Fault Earthquake Tsunamis.* National Oceanic and Atmospheric Administration. Retrieved from: https://www.dnr.wa.gov/Publications/ger_ofr2009-9_tsunami_hazard_tacoma.pdf. March 29, 2021.

Watters RJ, Zimbelman DR, Bowman SD, Crowley JK (2000) Rock mass strength assessment and significance to edifice stability, Mount Rainier and Mount Hood, Cascade Range volcanoes. *Pure and Applied Geophysics* 157: 957–976.

Wells RE, McCaffrey R (2013) Steady rotation of the Cascade arc. *Geology* 41: 1027–1030.

3 Water

In most cases, geology is the most powerful force in building the physical environment. However, in the Puget Sound, water in the form of ice has also made an enormous contribution. The massive glaciers scoured the land surface, and as they retreated, their melt water continued to erode the surface. Even after the last Ice Age melted away, water continued to sculpt the land. The Sound occupies a trough between the Cascade and Olympic Mountains. The water that feeds the Puget Sound and the surrounding region comes from several sources, including precipitation, river flow, residential and industrial releases, and of course, the Pacific Ocean. The Sound is tidal, and so, the Pacific has great influence on the Sound.

OCEAN WATER

PACIFIC OCEAN

The Puget Sound is a large salt-water estuary. It is open to the Pacific Ocean at three points. The main connection is through the Strait of Juan de Fuca or Admiralty Inlet, and there are two minor connections at Deception Pass and Swinomish Channel. The Sound is one part of several bodies of water that are collectively known as the Salish Sea. The other parts are the Strait of Juan de Fuca and the Strait of Georgia, which separates the island of Vancouver from the mainland of British Columbia.

TSUNAMIS

Tsunamis are powerful forces, but the Puget Sound is relatively protected from tsunamis from the Pacific Ocean. However, it is vulnerable to those from within the Sound generated by landslides and earthquakes. Landslides are the most common cause of tsunamis in the Sound. For example, an earthquake in 1949 near Olympia caused a landslide in the Tacoma Narrows that caused a tsunami of 2–3 m. The most dangerous source of tsunamis in the Sound is an earthquake on the Seattle fault. Atwater and Moore (1992) document a tsunami that occurred in the Puget Sound 1000–1100 years ago and speculate that an earthquake on the Seattle fault that occurred 500–1700 years ago. They found deposits of sand left by a tsunami at two sites north of the fault. Local tsunamis are very dangerous because they can occur with very little warning. Koshimura et al. (2002) also found evidence of a tsunami at that same time.

DOI: 10.1201/9780429487439-3

Diatoms (Bacillariophyta) are valuable in looking for tsunami histories. Their hard silicate remains can be easily found in the sand sheets that cover the land nearby and classified to determine whether the particular species lives in fresh or saltwater. Hemphill-Haley (1996) examined diatoms in the Puget Sound area and found that a tsunami occurred near Cultus Bay and West Point about 300 years ago.

Walsh et al. (2003) developed a model to predict the areas that would be inundated by a tsunami caused by an earthquake on the Seattle fault. This fault is certainly capable of generating a significant tsunami. A dead tree in the tsunami deposit from about 1000 years ago was matched to a drowned forest in Lake Washington. The cause of the tsunami was a landslide from Mercer Island. In addition, a sand layer along multiple streams from the Snohomish delta was likely left by a tsunami generated by an earthquake in AD 900–930. The map created by Walsh et al. (2003) predicted flooding of large areas of Seattle.

Sea Level Rise and Fall

Sea levels have risen and fallen many times as the Earth experienced Ice Ages in which massive amounts of water were frozen in the ice caps and glaciers that covered much of the Northern Hemisphere. In the late Pliocene about 2.67 million years ago during the last Ice Age, much of North America was covered with vast ice sheets. In fact, about a quarter of the Earth's continents were covered in ice. Later as the ice melted, the sea levels rose to cover more of the coastline. The water was more than 60 m above the current levels. For the next several hundred thousand years, sea levels rose and fell as the Ice Ages cycled. About 15,000 years ago, the last cold period ended and the ice began to melt. Sea levels rose some 90–120 m.

Glaciers

Over the last millions of years, sea levels have risen and fallen as the Earth has warmed and cooled and the ice masses at the poles and in glaciers have waxed and waned. The exact timings are not known (Otvos, 2014). Nevertheless, coastal regions were exposed as sea levels drop and inundated as they rise. In addition, other factors can affect sea levels in local areas (Shugar et al., 2014). Crustal deformation occurs during subduction. As plate movements increase subduction, the land rises and the relative sea level is lowered. When the stress is released by an earthquake, the land subsides and the relative sea level rises. Isostasy refers to the depression of the land by the sheer weight of ice on it. Land near the edge of the ice also rises. As the ice melts, the weight is removed and the land rises. Sediments can be compacted so that relative sea levels rise, or new sediments can fill in the water so that relative sea levels decrease.

The Earth has experienced at least five major ice ages. The last Ice Age started about 2.5 million years ago and involved at least seven advances and retreats of ice from Canada. The last major glaciation is known as the Cordilleran

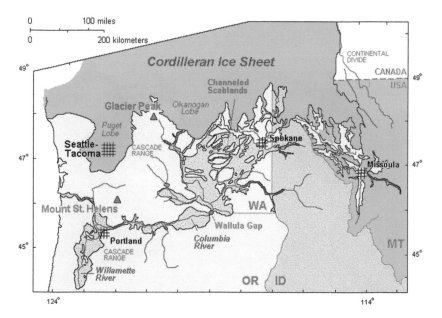

FIGURE 3.1 Cordilleran Ice Sheet. In the last Ice Age, the Cordilleran ice sheet covered much of North America. The masses of ice scoured the landscape and later, the melting ice caused more damage from erosion (Map courtesy of the US Geological Survey).

Ice Sheet, and it lasted from 18,000 to 30,000 years ago (Figure 3.1). In that period, glaciers slowly migrated southward, and by 17,400 years ago, it reached south of Olympia. By the Last Glacial Maximum, the ice over present-day Seattle was 1000 m thick. It did not last long and had retreated 100 km north of Seattle within 200–300 years.

The Cordilleran Ice Sheet redefined the major geologic features of the region. It carved out Puget Sound as it ground over the terrain. Before that, the Sound had been a river valley. With all of that ice, sea levels were much lower than today. The weight of the ice compressed the ground under it. The ground in Pioneer Square in Seattle was nearly 85 m lower than it is today. Once the glacier retreated, the ground bounced back. The ice advanced about 135 m per year. The ice at the front of the glacier melted and the water pushed sand, soil, and gravel in advance of the ice (Figure 3.2).

Sedimentation rates increase dramatically during glaciation events. Much of the Puget lowland is covered with sediments several hundred meters deep (Booth et al., 2003). In some areas, they exceed a kilometer below modern sea level and down to the bedrock. The Vashon Drift, as they are the sediments are known, can be divided into several units. Advance deposits include lacustrine silt and clay from proglacial lakes. Outwash includes sand and gravel from streams from the melting ice. Till is unsorted sand, gravel, silt, and clay from beneath the ice. Ice

FIGURE 3.2 Hoh Glacier on the Olympic Peninsula (Photograph courtesy of the National Park Service).

and ice-marginal deposits include various materials deposited on or near the ice. Finally, recessional deposits are well-sorted sand and gravel left by streams from the melting ice. In the northern and central parts of the Puget Sound, the sediments include glaciomarine drift that was left as the ice deposited material into the saltwater regions. The outwash, advance, and recessional deposits are quite permeable, but lacustrine and till are much less so. Permeability determines the path of water moving through the ground. Even old buried river beds can provide a channel for water to move underground. As the material consolidates, it becomes more resistant to erosion.

Additional processes related to the glaciers (e.g., erosion, landslides, subsidence, earthquakes) have continued to redefine the post-glacial terrain around the Sound. The most powerful was the Oceola mudflow from Mount Rainier 5700 years ago (covered in detail in Chapter 2).

Much of the evidence of the rise and fall of sea levels around the Puget Sound has been obliterated by the action of the glaciers. The receding glacier left behind many changes to the environment. The melting ice produced streams. Large blocks of ice falling from the glacier became buried in the sediment and left depressions called kettles. Others formed ponds. The retreating ice also dammed up water. For example, Glacial Lake Carbon was created by damming the Carbon River. Around 16,850 years ago, the ice dam broke to release a massive flood that covered much of Thurston and Pierce Counties and parts of Lewis and Grays Harbor Counties.

FRESH WATER

RAIN AND ATMOSPHERIC RIVERS (FIGURE 3.3)

The Puget Sound region is well known for its rain. Typically, in mid-October, a low-pressure cell, called the Aleutian Low, becomes stronger and moves to the southeast over the Aleutians and the Gulf of Alaska (Sound Science, 2007). The winds blow counterclockwise. Winds in a high-pressure system off Southern California blow in a clockwise manner. Those two systems combine to channel moist air into the Puget Sound region throughout the winter months.

The Cascade and Olympic Mountains, especially the west or windward slopes receive large amounts of rainfall and snow each year. The rainfall is related to orthographic lifting in which water-laden air is forced to higher altitudes by the terrain. As the air rises and cools, clouds form and rain or snow begins. They can get over 500 cm of rain per year, and the snow on Mount Rainier averages over 4 m. The Puget Sound lowlands receive less, but still have rainy winters.

There are variations in this pattern. For example, about one year in four involves an El Nino event in which the Aleutian Low moves farther south so that

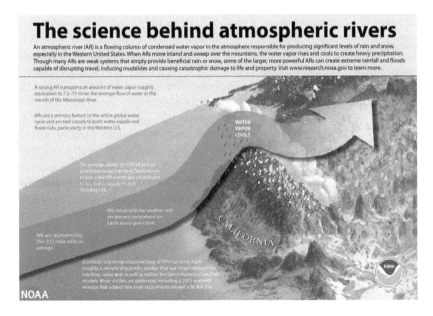

FIGURE 3.3 Atmospheric rivers (Diagram courtesy of the National Oceanic and Atmospheric Administration).

the storm track hits Northern California rather than Washington and Oregon. There is also the Pacific Decadal Oscillation that works on a cycle of 20–30 years. In these cases, the Aleutian Low is exceptionally strong or weak and thus results in somewhat warmer or cooler periods in the Puget Sound region. Global warming will likely change these patterns again.

On January 12, 2021, a massive storm struck the Puget Sound region (Cappucci, 2021). After an already wet couple of weeks, Olympia received another 5.0 cm of rain and Tacoma over 7 cm. Snow in the mountains totaled several meters. The air arriving in the region was carrying 760 million gallons of water every hour. The storm was a Level 5 atmospheric river (on a scale of 1–5). On average, only two storms of this magnitude hit the West Coast each year. In addition to the rain, heavy winds also buffeted the area. The year 2010 also featured strong atmospheric rivers (Ralph and Dettinger, 2011).

Washington State is well known for its rain. Its temperate maritime climate features wet winters and relatively dry summers. Most of the precipitation occurs from late October to mid-March. The annual rainfall varies. The mountains get up to 2000–5000 mm per year (Neiman et al., 2011). Those same mountains influence the rain that other areas receive. Seattle gets about 950 mm, but some places get less than half of that.

The river banks in Western Washington typically can handle a lot of water. The terrain includes steep and deep valleys. However, even they can be overwhelmed at times. The arrival of atmospheric rivers is one of those events. These large-scale precipitation events involve 100–150 mm of rain in a 24-hour period, and seem to occur about every 2 years (Cordeira et al., 2019). However, those numbers can be much greater, up to 600 mm in some cases. These storms are caused by low-level jets along the front of warm sectors of winter cyclones in the eastern North Pacific (Dettinger et al., 2011). They carry very large amounts of water vapor in a long narrow path. They are often more than 2000 km long, but only a few hundred km wide, and they are concentrated in the lowest 2.5 km of the atmosphere. Yet they transport more than 90% of the water vapor in the midlatitudes. On satellite maps, they stretch from Washington back to Hawaii.

The US Army Corps of Engineers and the National Oceanic and Atmospheric Administration and other state and local officials attempt to manage water flow after major rainstorm to limit flooding. They open or close floodgates on the dams to allow water to flow as safely as possible. The limits on this control involve the rate and timing of rainfall and the fact that many streams are not controlled by dams.

Every major flood in the Pacific Northwest has been associated with an atmospheric river event (Warner et al., 2017; Barth et al., 2016). Examples occurred in February 1996, October 2003, November 2006, December 2007, and January 2009. The storm on November 6–7, 2006, was especially intense. The atmospheric river stretches from the eastern Pacific to Washington. Rainfall in the mountains was 250–750 mm resulted in much higher than normal run-off. Warner et al. (2015) reported that, at the end of the 21st century, global warming will increase the number of extreme storms that will hit the Pacific Northwest by 250%. Furthermore, those storms will likely strike earlier in the year in October

(Warner and Mass, 2017). Konrad and Dettinger (2017) point out that the intensity of an atmospheric river is only one factor in predicting flooding and damage. The storm can be less intense, but if it lingers for more than a day, the total rainfall can produce flooding. Damage from water is not the only effect of atmospheric rivers. Waliser and Guan (2017) showed that such a storm is accompanied by strong winds that can affect storm surge, wind hazards, wave heights, and coastal flooding.

RIVERS

The Puget Sound watershed is over 30,000 km^2 (USDA, nd) (Figure 3.4). None of the rivers is very long because the crest of the Olympic and Cascade Mountains defines the edges of the watershed. A number of rivers flow into the Sound, including the Cedar/Lake Washington canal, Duwamish/Green, Elwha, Nisqually, Nooksack, Puyallup, Skagit, Skokomish, Snohomish, and Stillaguamish Rivers. These account for 80% of the freshwater flowing into the Sound. In addition, a number of smaller streams flow directly into the Sound.

AQUIFER

The Puget Sound lowlands overlie a large aquifer that is in the forearc basin of the subduction zone between the Juan de Fuca and North American plates

FIGURE 3.4 Puget Sound watershed. The watershed includes several major rivers (Image courtesy of the US Geological Survey).

(Gibson et al., 2018). The upper layer of the aquifer is unconfined to partially confined, and the lower layer is confined and below 45 m. The upper layer is used for water, but the lower is not.

It is contained below by Tertiary and older rock units and to the sides by the glacial drift of the Fraser Glaciation (Jones, 1999). It contains deposits from rivers, glaciers, and interglacial periods. At least, four glaciations and their interglacial periods were involved in the process. Jones (1999) divided the whole entity into several subunits, including the alluvial valley aquifers, the surficial semi-confining unit, the Fraser aquifer, the confining unit, and the Puget aquifer. Vaccaro et al. (1998) described the flow through the aquifer and the recharge rates for the various units.

DROUGHTS

While the Puget Sound region is well-known for rain, it actually experiences droughts at time. They are usually associated with El Niño events and the Pacific Decadal Oscillation (Newton et al., 2003). Droughts reduce freshwater flow into the Sound with commensurate effects on the water quality. In 2000–2001, the second-worst drought in the history of the state occurred. It was the driest since 1976–1977 and among the five worst of the past 100 years. River flow was down 28–72%. Newton et al. (2003) found that the changes to the water chemistry affected the plants and animals in the Sound. Droughts also occurred in 1987–1989 and 2004–2005 (Skukla et al., 2011). The region is susceptible to droughts because it depends on the snowpack in winter to supply water in the dry summers. If the winter is dry or warm, the amount of snow is reduced. Droughts also affect the wildlife and particularly the migrating salmon (LWCSW, nd).

EFFECTS OF WATER ON LAND

LANDSLIDES

Landslides are an agent of major change in mountainous areas, even greater than erosion, and either shallow or deep seated, they are a particular hazard. Wild fires that clear off the vegetation leave hills more vulnerable without roots to add stability to the soil on slopes. That risk is enhanced if the ground has been saturated by previous rain.

Landslides are common in the Puget Sound region. The steep terrain increases the susceptibility to landslides. However, somewhat surprisingly, shallow earth slides after heavy rains are the most common (Baum et al., 2005). Rotational (movement along a curved rupture surface) and translational (movement along parallel planes) landslides also occur. The winters of 1995–1996 and 1996–1997 were particularly damaging. Baum et al. (2005) developed models to map the probability of a landslide that combine history and slope stability to compute the hazard.

Slow-moving landslides are more likely in weak, fine-grained bedrock with low to moderate hillslope gradients (Scheingross et al., 2013). These are active hillslope mass failures with nonturbulent downslope movement of hillslope material. The rate of movement is typically millimeters to meters per year. They can last for decades or centuries. Tong and Schmidt (2016) used InSAR data to study a land movement in a Cascade setting in the early winter of several years. Movement in the winter of 2009 and 2011 was the greatest and correlated with the local precipitation.

Severe winter storms result in more landslides. Handwerger et al. (2019) showed that storms increase the likelihood of landslides. Recently, the Washington state adopted ARkStorm, a statewide emergency planning scenario for extreme storms. Wills et al. (2014) combined information from ARkStorm with data on areas of greatest susceptibility to landslides to produce a model that can better predict where landslides might occur. Drought tends to reduce the number of landslides (Bennett et al., 2016).

COASTAL EROSION

The many steep bluffs in the region were created by the glaciers that scoured the land tens of thousands of years ago (Ecology, nd). Most of the shoreline around the Puget Sound is susceptible to erosion. Those actions change the shape of the coast and create beaches. In response, humans have constructed bulkheads and seawalls to slow the erosion. They can help, but they also can cause the loss of beaches and important habitats.

The Puget Sound has 4000 km of beaches (Shipman, 2010), and about half of the shoreline consists of bluffs. The beaches were made by erosion of material from the coasts. Some of the bluffs surrounding the Sound are known as feeder bluffs, and they are the ones that are slowly eroding to provide sand and gravel to the beach. The rate of erosion can vary greatly, and it is influenced by the height of the bluff, the proportion of sediment in the bluff, and the rate of erosion or retreat. Much of the sediment along the bluffs was leftover from the glaciers. It includes lake-bed clays, outwash sands and gravels, coarse till and marine drift, and material deposited by water flowing under the glaciers. The eroded material is then moved around by wave action until it finds a location where it becomes trapped, at least temporarily.

Over the last millions of years, sea levels have risen and fallen. As the sea levels rise from global warming, the foot of the slope is exposed to greater wave action, and the rate of erosion increases. Beaches come and go as the amount of eroded material increases or is lost. In the Puget Sound, the fetch is quite limited, and so, wave action is similarly limited. Longshore transport becomes more important for beach evolution. Rivers also carry sediment to the coast. They build deltas at their mouths. These combine with the eroded material to build beaches.

CONCLUSION

Water is not as powerful as a force as plate tectonics. Nevertheless, water has had an enormous effect on the Puget Sound region. It did not build the basic structure of the Sound, but it refined the shores and mountains and provided material for the land. During the last Ice Age, the massive glaciers of the Cordilleran Ice Sheet ground down many land features, and since then, erosion has continued to modify the surface in the Puget Sound region.

REFERENCES

Atwater BF, Moore AL (1992) A tsunami about 1000 years ago in Puget Sound, Washington. *Science* 258: 1614–1617.

Barth NA, Villarini G, Nayak MA, White K (2016) Mixed populations and annual flood frequency estimates in the western United States: The role of atmospheric rivers. *Water Resources Research* 53: 257–269.

Baum RL, Coe JA, Godt JW, Harp EL, Reid ME, Savage WZ, Schulz WH, Brien DL Chleborad AF, McKenna JP, Michael JA (2005) Regional landslide-hazard assessment for Seattle, Washington, USA. *Landslides* 2: 266–279.

Bennett GL, Roering JJ, Mackey BH, Handwerger AL, Schmidt DA, Guillod BP (2016) Historic drought puts the brakes on earthflows in Northern California. *Geophysical Research Letters* 43: 5725–5731.

Booth DB, Haugerud RA, Troost KG (2003) The geology of Puget Lowland rivers. In: Montgomery DR, Bolton S, Booth DB, Wall L (editors). *Restoration of Puget Sound Rivers*, pp. 14–45. University of Washington Press, Seattle, WA.

Cappucci M (2021) *Category 5 Atmospheric River Blasts Pacific Northwest, with up to 10 Inches of Rain Possible*. Washington Post (January 12, 2021). Retrieved from: https://www.washingtonpost.com/weather/2021/01/12/atmospheric-river-pacific-northwest/. March 22, 2021.

Cordeira JM, Stock J, Dettinger MD, Young AM, Kalansky JF, Ralph FM (2019) A 142-yearclimatology of northern California landslides and atmospheric riversv. *American Meteorological Society*, 1499–1510.

Dettinger MD, Ralph FM, Das T, Neiman PJ, Cayan DR (2011) Atmospheric rivers, floods and the water resources of California. *Water* 3: 445–478.

Ecology (nd) *Preventing Puget Sound Shoreline Erosion*. Department of Ecology, Washington State. Retrieved from: https://ecology.wa.gov/Water-Shorelines/Puget-Sound/Helping-Puget-Sound/Preventing-erosion. March 27, 2021.

Gibson MT, Campana ME, Nazy D (2018) Estimating aquifer storage and recovery (ASR) regional and local suitability: A case study in Washington State, USA. *Hydrology* 5: 7.

Handwerger AL, Fielding EJ, Huang M-H, Bennett GL, Liang C, Schulz WH (2019) Widespread initiation, reactivation, and acceleration of landslides in the northern California Coast Ranges due to extreme rainfall. *Journal Geophysical Research: Earth Surface*, 124.

Hemphill-Haley E (1996) Diatoms as an aid in identifying late Holocene tsunami deposits. *The Holocene* 6: 439–448.

Jones MA (1999) *Geologic Framework for the Puget Sound Aquifer System, Washington and British Columbia*. US Geological Survey, Washington. Professional Paper 1424-C.

Konrad CP, Dettinger MD (2017) Flood runoff in relation to water vapor transport by atmospheric rivers over the western United States, 1949–2015. *Geophysical Research Letters* 44: 11,456–11,462.

Koshimura S, Mofjeld HO, Moore A (2002) Modeling the 1100 bp paloetsunami in Puget Sound, Washington. *Geophysical Research Letters* 29: 9.

LWCSW (nd) *Drought Conditions and Local Salmon Populations – A Glimpse of the Future?* Lake Washington-Cedar-Sammamish Watershed. Retrieved from: https://www.govlink.org/watersheds/8/committees/1510/1510_5099_WRIA8_ILAFactSht_DROUGHT-150.pdf. March 22, 2021.

Neiman PJ, Schick LJ, Ralph FM, Hughes M, Wick GA (2011) Flooding in Western Washington: The connection to atmospheric rivers. *Journal of Hydrometeorology* 12: 1337–1358.

Newton JA, Siegel E, Albertson SL (2003) Oceanographic changes in Puget Sound and the Strait of Juan de Fuca during the 2000–01 drought. *Canadian Water Resources Journal* 28: 715–728.

Otvos EG (2014) The Last Interglacial Stage: Definitions and marine highstand, North America and Eurasia. *Quaternary International* 383: 158–173.

Ralph FM, Dettinger MD (2011) Historical and national perspectives on extreme west coast precipitation associated with atmospheric rivers during December 2010. *Bulletin of the American Meteorological Society* 93: 783–790.

Scheingross JS, Minchew BM, Mackey BH, Simons M, Lamb MP, Hensley S (2013) Fault-zone controls on the spatial distribution of slow-moving landslides. *GSA Bulletin* 125: 473–489.

Shipman H (2010) The geomorphic setting of Puget Sound: Implications for shoreline erosion and the impacts of erosion control structures. In: Shipman H, Dethier MN, Gelfenbaum G, Fresh KL, Dinicola RS (editors). *Puget Sound Shorelines and the Impacts of Armoring—Proceedings of a State of the Science Workshop*, pp. 19–34. US Geological Survey, Washington.

Shugar DH, Walker Ian J, Lian Olav B, Eamer Jordan BR, Neudorf C, McLaren D, Fedje Daryl (2014) *Post-glacial Sea-Level Change Along the Pacific Coast of North America.* SIAS Faculty Publications. Retrieved from: https://digitalcommons.tacoma.uw.edu/ias_pub/339.

Skukla S, Steinemann AC, Lettenmaier DP (2011) Drought monitoring for Washington State: Indicators and applications. *Journal of Hydrometeorology* 12: 66–83.

Sound Science (2007) *Puget Sound's Climate.* Encyclopedia of the Puget Sound. Retrieved from: https://www.eopugetsound.org/articles/puget-sounds-climate. March 22, 2021.

Tong X, Schmidt D (2016) Active movement of the Cascade landslide complex in Washington from a coherence-based InSAR time series method. *Remote Sensing of the Environment* 186: 405–415.

USDA (nd) *National Geospatial Center of Excellence.* US Department of Agriculture. Retrieved from: https://www.nrcs.usda.gov/wps/portal/nrcs/main/national/ngce/. March 28, 2021.

Vaccaro JJ, Hansen Jr AJ, Jones MA (1998) *Hydrogeologic Framework of the Puget Sound Aquifer System, Washington and British Columbia.* US Geological Survey, Washington.

Waliser D, Guan B (2017) Extreme winds and precipitation during landfall of atmospheric rivers. *Nature Geoscience* 10: 179–183.

Walsh TJ, Titov VV, Venturato AJ, Mofjeld HO, Gonzales FI (2003) *Tsunami Hazard Map of the Elliott Bay Area, Seattle, Washington: Modeled Tsunami Inundation from a Seattle Fault Earthquake*. Division of Geology and Earth Resources, Open File Report 2003–14. Washington.

Warner MD, Mass CF (2017) Changes in the climatology, structure, and seasonality of Northeast Pacific atmospheric rivers in CMIP5 climate simulations. *Journal of Hydrometeorology* 18: 2131–2141.

Warner MD, Mass CF, Salathé JR (2015) Changes in winter atmospheric rivers along the North American West Coast in CMIP5 climate models. *Journal of Hydrometeorology* 16: 118–128.

Warner M, Mass C, Shaffer K, Brettman K (2017) *Atmospheric Rivers, Climate Change, and the Howard A. Hanson Dam*. American Geophysical Union, Fall Meeting 2017. Retrieved from: http://cw3e.ucsd.edu/IARC_2018/ARs_and_Climate_Variability/Warner_IARC2018.pdf. March 22, 2021.

Wills C, Perez F, Branum D (2014) New method for estimating landslide losses from major winter storms in California and application to the ARkStorm scenario. *Natural Hazards Review*. DOI 10.1061/(ASCE)NH.1527-6996.0000142.

4 Geomorphology of Puget Sound

INTRODUCTION

Geomorphology is the science of how the surface topography of the earth is a result primarily from the interactions of the environment upon the underlying geology (Hunt, 1988). The resulting landscapes reflect not only the chemical and physical composition of the underlying rocks, but also the effect of water and weathering of those rocks, atmosphere, and hydrosphere; the soils created from the rocks by the presence and flow of water; the microbes in the soils; the botanical diversities that are successful in the environment; and the combined influence of herbivores and predators within the ecological community, namely the biosphere. Finally, the effect of humans also may contribute significantly to the landscape and thereby may undermine some of the balancing forces that had been in effect for millennia.

This chapter describes a few examples of the geomorphology of regions within the Puget Sound region that represent the results of different interactions between the environment, humans, and the earth's surface and which have created the landscapes we see today. In this and other chapters, we will be following the lead of Kruckeberg when referring to Puget Sound or the Puget Sound region:

> "Sometimes the terms "Puget Sound" and "Puget Sound and adjacent waters" are used for not only Puget Sound proper but also for waters to the north, such as Bellingham Bay and the San Juan Islands region (Kruckeberg, 1991).

In order to encompass the region in a more rigorous manner, we will also include the eastern Olympic Peninsula, eastern Vancouver Island, as well as the western foothills of the Cascades. This area roughly approximates to a circle of 100 km radius centered on Port Townsend, Washington.

In this chapter, we summarize the eight processes that affect geomorphology as they relate to the Puget Sound region: aeolian, biological, fluvial, glacial, hillslope, igneous, tectonic, and marine. We will describe a number of localities in the Puget Sound region that demonstrate either a particular process or a combination of processes. These localities are of interest when viewed in the context of the Puget Sound region, including how the underlying rock influences the overlying vegetation and how past volcanism and plate tectonics result in the structures of the Olympics and Cascades and the Puget Lowland. We also

DOI: 10.1201/9780429487439-4

summarize how precipitation and water flow, estimated over the past five to eight million years, have affected the geomorphology of the Puget Sound region.

Geomorphological processes generally fall into three groups: (1) the production of rock and mineral fragments (regolith) by weathering and erosion, (2) the transport of that material, and (3) its eventual deposition and interaction with the underlying surface (Derbyshire et al., 1979). The processes are further categorized as follows.

AEOLIAN

Named for Aeolus, the Greek god of the wind, these are wind-generated geologic processes. Wind is characterized as the movement of air between atmospheric high-pressure and low-pressure systems, in particular at the interface between the atmosphere and the surface of the Earth. A greater difference in pressure between the two systems results in a greater velocity of the wind. Depending upon the wind velocity, there will be a number of different effects that the wind may have upon the underlying rocks, sand, silts, minerals, and the biosphere. Silts and sands are generally defined according to size, silts being smaller than 0.0625 mm down to 0.004 mm, and sands being larger than 0.0625 mm up to 2 mm (ASTM, 1985). The effects can be transport, erosion, and deposition of rocks, sand, and minerals as well as the consequences upon vegetation and precipitation.

For wind velocities less than 10 km/h (~ 5 m/s at 1 m above the surface) sand particles remain in place (Sloss et al., 2012); however, at greater wind velocities (≥5 m/s at 2.4 m above the surface) sand particles may be transported many tens or hundreds of meters; in addition, greater wind velocities can transport proportionally more sand particles than at lower velocities (Webb et al., 2016). Coarse sand particles, usually a feldspar or quartz, both high in silica, having more abrasive properties than an equivalent softer rock (e.g., gypsum, apatite, or other minerals that make up much of sedimentary rocks), are particularly important in the process of erosion, whereby the force of the sand particle upon a rock or mineral surface is sufficient to break off fragments of the rock or mineral, either to be carried further by the wind or deposited in the vicinity, where the fragments may later be subjected to fluvial action (see "Fluvial" below). Interestingly, vegetation cover can also influence the local wind velocity and erosion and deposition rates (Webb et al., 2016).

BIOLOGICAL

Biogeomorphological processes, perhaps unexpectedly to the layperson, have probably greater influence upon geomorphology of the land and the marine or freshwater environments than any other processes. First and foremost are those processes relating to the underlying geology, the subsequent weathering of those rocks by aeolian processes to create a fragmentary mineral layer, and the microbes and plants that use those minerals to grow and propagate; their waste products add to the biological and chemical detritus that, combined with the fragmentary minerals and chemical weathering, form the soil. The soil depends upon (1) the type of minerals released from the underlying rock, (2) the environmental conditions, including heat/cold, wet/dry conditions, and oxygen availability, and (3) the resulting ecosystem that is supported by that particular soil and environment.

An example of how the biota of a location can modify landform changes are the plants and animals present in marshlands, wetlands, or estuaries, for example, eelgrass (*Zostera marina*), saltgrass (*Distichlis spicata*), and Lyngbye's sedge (*Carex lyngbyei*) and Olympia oysters (*Ostrea lurida*). Their presence plays an important role in slowing and reducing the force of tides and storm-waters, thereby slowing the rate and effects of erosion by marine geomorphological processes. Pickleweed (*Salicornia virginica*), a succulent, is considered a "pioneer" species that is an early inhabitant of tidal flats in low sandy marshes (Seliskar and Gallagher, 1983).

Second, zoogeomorphology, whereby animals influence the form of the land is an important process. Examples include beaver dams that modify the flow of water and sediment across the land, the action of burrowing animals, digging for tubers and roots, the formation of nutrient-rich environments left by the roots and above-ground parts of trees and shrubs both before and after death, can affect the way that the soil components are transferred from one layer to another, thereby providing new environmental conditions for organisms to take advantage of. Another important mammal that has a profound effect on the composition of the vegetation in waterways is the muskrat (*Ondatra zibethicus*); by selectively removing preferred plant species it leads to changes in the abundance and distribution of wetland plant ecosystems (O'Neil, 1949; van der Valk, 1989; Keddy, 2010).

Notably absent from marshes and coastal wetlands of the Pacific Northwest are decapods (crabs and their allies), gastropods, and bivalves, when compared to Atlantic coastal marshes (Seliskar and Gallagher, 1983). Marshes also serve as important nurseries for juvenile fish, in particular salmonids (*Oncorhyncus* spp.), anchovies (*Engraulis mordax*), and perch (*Cymatogaster aggregata*) (Seliskar and Gallagher, 1983).

Third, and perhaps less obvious, is that the combined ecosystem can influence the balance of the atmospheric carbon dioxide, which ultimately can modulate the climate. One notable exception to the large influence of biological activity upon the geomorphology of a region is to be found in Antarctica, but of course, this does not relate to the contemporary Puget Sound region.

FLUVIAL

Fluvial refers to any moving or stationary water body on the terrestrial landscape, such as creeks, streams, and rivers. Beginning from a spring or as meltwater from a glacier or snow, the water seeks the lower altitudes on its course its destination, either a lake or the sea. As it flows, it will erode the enveloping stream-bank and its bed, forming a V-shaped valley; the amount of erosion will depend upon its velocity downhill. Greater velocities result in increased erosion, which translates into more geomorphological variation over time; fast moving streams usually have a rather straighter path than those slower-moving waters but will deposit the sediment when suddenly slowed; alluvial fans on the sides of mountainous cliffs are an example of such flow and deposition. As the terrain becomes less steep, the water velocity slows, thus reducing the amount of bank and bed erosion and the path of the river may meander. Greater velocities up-stream also result in the amount of sediment that may be transported and thus large volumes of sediment may be deposited as the river slows to a meandering phase. This in turn may build up the surface of the land through which the river flows, and may result in changes in the ecosystems and landscape. The avail-ability of new nutrients may encourage other species to proliferate in that en-vironment and the slow pace and slower erosional rate may result in the formation of ox-bow lakes and fluvial terraces, a botanically diverse river system that provides additional biological niches for other animals and plants.

The underlying topography of the Puget Sound area, the mix of volcanos, highlands, valleys, and flood plains are all impacted by the rivers and streams which originate in the ranges to the east, in particular Mount Rainier (4394 m), Mount Baker (3287 m), Glacier Peak (3214 m), and from the west, the Olympic Mountains (>2000 m). Since the Cascades reached their present elevations over 2 mya, weather has been a constant source of erosion, and thus the amount of uplift equals the amount of ongoing fluvial erosion, resulting in the ranges remaining at an almost constant elevation (Reiners et al., 2003; NSF, 2003).

GLACIAL

The past effect of glaciers is a significant process within the geomorphology of the Puget Sound region. The movement of glaciers down an old river valley slowly erodes the valley sides and the floor producing rock debris and creating a U-shaped valley; when the climate warms up, the glacier melts and retreats up the valley, leaving the rock debris on the surface, termed moraine. The Puget Sound region was subjected to direct effects of glaciation during the last ice age, particularly the Vashon Glaciation (20 to 15.2 kya) (kya: thousand years ago) and the Younger Dryas cooling (13 to 11.1 kya); during the peak of this last ice age meltwater from under the overlaying ice sheet scoured the major trough that is Puget Sound, re-sulting in the glacially derived sedimentary outcrops exposed along the rivers and coastal bluffs seen today (Burns, 1985; Booth, 1994; Shipman, 2010).

Another glacial feature seen in the Puget Sound region is kettles; they are small depressions found in the glacial sediment. These are formed when blocks of ice are left behind as the glacier retreats, glacial sediment then covers these ice blocks, and when the ice eventually melts, the overlying sediment clumps down into the now-empty space resulting in a depression (WSDNR, 2021a).

The southern Puget Lowland has some mysterious large landscape features called Mima Mounds. They are generally circular, chest-high dome-shaped, and composed of a dark gravelly soil (WSDNR, 2021b). They are somewhat evenly distributed across the landscape and spaced at similar distances from one another, although size (radius) varies somewhat. Their origins are uncertain but as most of them lie within recessional glacial outwash deposits they may be the result of an unknown geomorphic process. Recently, Johnson and Johnson (2012) concluded that they arose from a polygenic process, comprising bioturbation, seasonal frost action, and erosion processes, and with occasional aeolian inputs, best accounted for the mounds, their well-sorted stone borders, and the poorly sorted rubbly soil intermound pavements (Johnson and Johnson, 2012).

Hillslope

Soil, eroded minerals, sand, and rock will move down a slope as creep and accumulate at the base in the valley; the slope surface can be anything from essential vertical to almost flat and the angle will determine the rate of creep. The moment that creep begins depends upon (1) the rate of weathering of the rock; (2) the amount of water present within the soils; and (3) the composition of the underlying rock. Ongoing hillslope processes will change the topology of the hill's surface, resulting in a steadily increasing base height, and will further retard the rate of creep towards the valley floor. In addition, animal activity as reported above, may also affect the rate and onset of creep.

In the past, tectonic forces have contributed to crustal uplift, resulting in steep slopes—subsequent weathering by rain, snow, streams, and rockfalls, have re-sulted in sporadic landslides along the streams and coastline, which in turn cause changes in water movement, etc. Landslides are frequently common in the mountain ranges, for example along the upper Quinault River in the Olympic National Park (Gavin et al., 2013). More recently in Washington State, in years where there are both short-term, intense storms and above-average annual rainfall, additional factors influence the occurrence of hillslope erosion (land-slides). The first was new construction and development where the hillslopes are modified by cutting and filling. The second is that of the long-term dry period that began following the Second World War, coincident with the increase in development.

Igneous

Igneous processes, defined as the result of volcanic activity (both eruptive and in-trusive), may catastrophically alter the landscape, thereby re-setting the paths of flowing water, migration routes of birds and animals, and delay the establishment of the botanical ecosystem for many years. As mentioned in Chapter 2, five large active volcanoes are located in Washington State (Baker, Rainier, Mount St. Helens, Adams, and Glacier Peak), three of which are within the Puget Sound watershed, Rainier, Baker, and Glacier Peak (Figure 4.1); periodic eruptions from each of them has inundated the landscape with lava and pyroclastic flows, and lahars. In addition, much of the deeper underlying bedrock of the Puget Sound region originates from older pillow basalt terranes from submarine eruptions tens of millions of years ago.

FIGURE 4.1   Glacier Peak. Glacier Peak is another active volcano in the Puget Sound region. Since the end of the most recent ice age, Glacier Peak has erupted multiple times during at least six separate episodes. Glacier Peak and Mount St. Helens are the only volcanoes in the State that have had large explosive eruptions in the past 15,000 years (Photograph courtesy of the US Geological Survey).

TECTONIC

These processes are the result of the constant motion of the continental and oceanic plates, as described in Chapter 2. The resulting constant movement, which can be instantaneous or over hundreds of years, causes the relationship between adjacent eco/geomorphological systems to be in constant flux and can be observed as differences in vegetation and the animal life they support. Tectonic activity, such as earthquakes, may cause the upper portions of the crust to rise or collapse, which leads to more changes in the environment, fluvial flow, and weathering.

The Puget Sound Bay region is part of the Juan de Fuca and North American plate complex and so has been subjected to not only periodic deposition by marine sediments for more than 400 million years, but also is under constant seismic activity due to the eastward movement of the Juan de Fuca plate, and to a lesser extent, the northward movement of the adjacent Pacific plate. This has resulted in buckling of the underlying crust and subsequent weathering to create the characteristic mountainous Olympic Penninsula/Vancouver Island complex and North Cascades, the lowland Puget Sound/Salish Sea waterways, and the up-thrusted San Juan Islands (Brandon et al., 1988). The region is also being tilted in a north-south direction, the northern portion, including the San Juan Islands, is rising, and the southern portion, including southernmost Puget Lowlands, sinking. There have also been periodic local seismic subsidence and emergence events associated with Holocene faulting, which has affected the shorelines near the faults (Bucknam et al., 1992).

MARINE

Marine processes are those of the action of waves, marine currents, and seepage of fluids, including seawater mixed with decomposing organic material, through the seafloor. In the Puget Sound region, along with fluvial deposition from the region's many river systems, wave action has eroded the 3000–4000 km coastline and transported sediment, predominantly in a south to north direction, whereas tidal action generally follows a northwest to southeast direction, from the Strait of Juan de Fuca to southern Puget Sound (Finlayson, 2006; Shipman, 2010). The combination of these two forces contributes to sediment transport along the shores of Whidbey Island and the Kitsap Peninsula. About half of the Sound's coastline consists of bluffs and small barriers; the remainder are bedrock shorelines, river estuaries, and large river deltas (Shipman, 2010). Storm and heavy surf action are also contributing to the erosion of many of the uplifted regions of the San Juan Islands (see Figure 10.1).

In addition to current marine erosion forces, sea-level changes during the past 15,000 years have also influenced the shoreline and estuarine topography. Combined with changes in isostatic rebound of the land following the removal of the ice and local tectonic movements, has led to differential rates of subsidence in southern Puget Sound compared with those in the north (Shipman, 2008). As mentioned above, the tectonic forces have resulted in uplift and folding of the underlying rock, which, followed by fluvial action, has resulted in large areas of marshland and lagoons protected behind sand bars that run along the coast. The marshlands contain brackish water whereas the lagoons are predominantly freshwater. The lagoons are an important resource for seabirds, particularly diving birds that must rinse away the seawater from their feathers after fishing.

OVERVIEW

The Puget Sound region comprises a wide range of topographies that are distinct to the locality. From a geomorphological viewpoint, the region may be reviewed within four distinct localities: to the south, Puget Sound and its adjacent riparian inflows; to the east, the foothills of the western Cascades; to the north, the San Juan Islands, Bellingham Bay, the eastern portion of the Strait of Juan de Fuca, and the southeastern portion of Vancouver Island; and to the west, the eastern portion of the Olympic Peninsula (Shipman, 2010). Figure 1.2 illustrates Puget Sound and the surrounding regions that are the focus of this book.

The majority of the landscape's form is due to the past and ongoing tectonic activity, the nature of the underlying bedrock, past glacial activity, as well as being influenced by the current climate and weather. Occasionally, during an earthquake, sea-level changes have been instantaneous (Bucknam et al., 1992; Sherrod et al., 2004).

The region is dominated by the Puget Lowland, a depression resulting from subduction of the Juan de Fuca plate beneath the North American plate and comprises Puget Sound, Camano and Whidbey Islands, the San Juan Islands, and Bellingham Bay. As described in Chapter 2, a large proportion of the western North American continent comprises ancient terranes, namely continental masses made up of ancient island arcs, mountains, and sea-beds that rode upon the Farallon plate and accreted at the western edge of the North American plate margin as the Farallon plate was being subducted. The Puget Lowland comprises this older bedrock, but is covered by Pleistocene sediments deposited from the weathering of the rising Cascades and Olympic Peninsula (Shipman, 2010); these sediments primarily are relatively recent depositions, deriving from the major rivers flowing from the Cascades and the Olympics during the Holocene (11.7 kya to the present), as well as being affected by eruptions and mudflows from the Cascade volcanoes. Much of the sediment found on beaches surrounding Puget Sound and its adjacent waterways derives from coastal bluff erosion, such as seen on Guemes Island (Bellingham Bay), where Pleistocene Epoch (90 to 100 kya) glacial and glaciofluvial sedimentary formations, made up of silts, gravels, and boulders, can reach up to 30–40 m in height (Kahle and Olsen, 1995; Shipman, 2010).

The shorelines of Puget Sound can be categorized into one of the following four geomorphic systems: (1) rocky coasts, typified by the San Juan Archipelago comprising igneous and sedimentary bedrock of Middle Jurassic to Early Cretaceous age (170– 100 mya); (2) beaches, including those with coastal bluffs and barrier beaches; (3) embayments, including estuaries and lagoons; and (4) large river deltas, including the large rivers that drain the North Cascade and Olympic mountain ranges (see Figure 4.2) (Shipman, 2008; GSA, 2018).

Pocket estuaries are river outflows whereby the mouth of the estuary is narrower than its main body of water. These have been particularly useful as sites for reintroducing non-natal Chinook salmon to the region's waterways as their waters are somewhat protected from wave and tidal action (Beamer et al., 2003).

One interesting aspect of the Puget Sound region is the presence of many barrier beaches. These form when wave and tidal action in a bay or estuary is limited and thus a barrier of sediment and rocks may build up across the mouth of the inlet; an example is a barrier estuary at the mouth of Stavis Creek on Hood Canal formed by spits of sand and other regolith (Shipman, 2008). Another type of barrier forms when the underlying rock resists erosion and tidal and wave action create a recurved spit that acts as a barrier to protect a lagoon; an example is seen at Point Monroe on the north end of Bainbridge Island (10 to 12 km west of Seattle) (Shipman, 2008).

FIGURE 4.2   Skagit River delta (Photograph courtesy of US Geological Survey).

Much of the Puget Lowlands, with the exception of the San Juan Islands and the Strait of Juan de Fuca, is covered by recent glacial sediments. The surrounding uplands, the Olympic and North Cascade ranges, are exposed bedrock. The Olympic Mountains are predominantly a mixture of sandstones, mudstones, minor conglomerate, siltstone, and basalt from the late Oligocene and Early Miocene (Hurricane Ridge thrust and Olympic structural complex); the Cascade ranges mainly comprise Eocene oceanic crust (Crescent Formation basalt and Siltez River Volcanics) (Tabor and Cady, 1978a, 1978b; Clowes et al., 1987).

An important contributor to landform changes in the past has been the combined effect of volcanism, glaciation, and fluvial actions upon the volcanic peaks. One example is that of lahars, produced by large debris avalanches during eruptions on the slopes and in the valleys descending from a volcano. These lahars are a by-product of hydrothermal alteration, whereby hot, sulfur-rich volcanic gases encounter groundwater. The sulfur gases dissolve into the groundwater to create sulfuric acid, which then attacks, leaches, and replaces the minerals in the rocks with weak and water-saturated clay minerals. The presence of clay mobilizes any collapsed material during earthquakes and/or volcanic eruptions, thus allowing it to flow as a liquid; nothing can stop this terrifying catastrophic event.

About 5600 years ago, a massive lahar tens of meters deep, the Osceola mudflow, originated from the Winthrop and Emmons Glaciers on Mount Rainier and flowed all the way to the site of present-day Tacoma and southern outskirts of Seattle (see Figure 2.3); most of the resulting crater on the flood-plain was infilled with subsequent lava eruptions (USGS, 1999, 2014). These also bring new minerals to the lower-lying plains and can result in changes in the vegetation cover.

Peat is another additional depositional element found in the Puget Sound region. Peat is composed of partially decomposed and disintegrated plant matter, particularly *Sphagnum* moss, sedges and grasses, and occurs in poorly drained low-lying areas. A typical example of peat bogs are found on Guemes Island, Whidbey Island, throughout the Puget Lowlands, in low-lying areas of the eastern Olympic Peninsula, and in montane glacial lakes of the North Cascades (Rigg, 1958; Shipman, 2010).

South and west of Seattle are natural largely fresh-water lagoons sourced by the flow of rivers and streams. They were formed as the rivers flowed over the lowlands that characterize Puget Sound area forming a shallow bay behind sand-gravel spits. Some examples include Burley Lagoon located on the Key Peninsula at the heart of Puget Sound, Kiket Lagoon and Lone Tree Lagoonare both in Skagit Bay, between Whidbey Island and the mainland, and Deer Lagoon, at the south end of Whidbey Island. All of these lagoons and marshes play a critical role in the biosphere of the Puget Sound area as source of fresh water for sea birds, particularly diving birds such as pelicans (*Pelicanus occidentalis* and *P. erythrorhynchos*, both rare) and cormorants (*Phalacrocorax penicillatus*, *P. pelagicus*, and *P. auritus*) to wash out the salt from their feathers.

To the north of the Puget Sound lies Boundary Bay, an intertidal coastal wetland, and which has a similar role. In addition, the estuary and the larger lagoons are key stopover points on the Pacific Flyway (Carlisle et al., 2009) and is an essential breeding, feeding, and nesting ground for over 370 species of migratory and native birds, including six endangered species, some of which include the western sandpiper (*Calidris mauri*), dunlin (*Calidris alpine pacifica*), Virginia rail (*Rallus limicola*), and song sparrow (*Melospiza melodia*) (Seliskar and Gallagher, 1983). Boundary Bay has been designated a Hemisphere Reserve by the Western Hemisphere Shorebird Reserve Network and it is also designated a Canadian Important Bird Area (WHSRN, 2008a, 2008b).

The lagoons, tidal marshes, estuaries, and beaches (including barrier beaches) also support a wide variety of other organisms, including phytoplankton, macroalgae, and polychaetes; eelgrass (*Zostera marina*); and fish, often juvenile marine fish, such as northern anchovy (*Engraulis mordax*), shiner perch (*Cymatogaster aggregata*), starry flounder (*Platichthys stellatus*), and surf smelt (*Hypomesus pretiosus*) (Seliskar and Gallagher, 1983).

Regarding how the climate, precipitation, and fluvial action have molded the landscape of the Puget Sound region, fossils of plants dating from the late Pliocene Epoch (~3 mya) suggest that the climate was cooler and moister than the earlier Miocene, as well as compared to today (Axelrod and Deméré, 1984;

Reiners et al., 2002). By the time of the Last Glacial Maximum (21 kya) the mean summer temperature was 6 to 7°C cooler than today and having 40% lower mean annual precipitation (Reidel, 2017). Fossil evidence also exists that indicates an extremely diverse fauna of large animals that fed on the vegetation supported by greater levels of precipitation (see Chapter 5).

REFERENCES

ASTM (1985) Classification of soils for engineering purposes: Annual book of ASTM Standards, D 2487-83. *American Society for Testing and Materials* 4: 395–408.

Axelrod DI, Deméré T (1984) A pliocene flora from Chula Vista, San Diego County, California. *Transactions of the San Diego Society Natural History* 20(15): 277–300.

Beamer E, McBride A, Henderson R, Wolf K (2003) *The Importance of Non-Natal Pocket Estuaries in Skagit Bay to Wild Chinook Salmon: An Emerging Priority for Restoration.* Skagit System Cooperative Research Department, LaConner, Washington.

Booth DB (1994) Glaciofluvial infilling and scour of the Puget Lowland, Washington, during ice-sheet glaciation. *Geology* 22: 695–698.

Brandon MT, Cowan DS, Vance JA (1988) *The Late Cretaceous San Juan Thrust System, San Juan Islands, Washington.* Special Paper 221, The Geological Society of America, Inc., Boulder, Colorado.

Bucknam RC, Hemphill-Haley E, Leopold EB (1992) Abrupt uplift within the past 1700 years at Southern Puget Sound, Washington. *Science* 258: 1611–1614.

Burns R (1985) *The Shape and Form of Puget Sound: Seattle, Wash.* Washington Sea Grant, University of Washington Press, Seattle, Washington.

Carlisle, JD, Skagen, SK, Kus, BE, Riper, CV, Paxtons, KL, Kelley, JF (2009) Landbird migration in the American west: Recent progress and future research directions. *Condor: Ornithological Applications* 111: 211–225. doi:10.1525/cond.2009.080096.

Clowes RM, Brandon MT, Green AC, Yorath CJ, Sutherland-Brown A, Kanasewich ER, Spencer C (1987) LITHOPROBE – southern Vancouver Island: Cenozoic subduction complex imaged by deep seismic reflections. *Canadian Journal of Earth Sciences* 24: 31–51.

Derbyshire E, Gregory KJ, Hails JR (1979) *Studies in Physical Geography.* Butterworth-Heinemann; Elsevier, Oxford, United Kingdom.

Finlayson D (2006) *The Geomorphology of Puget Sound Beaches: Seattle, Wash.* University of Washington, Seattle, Washington.

Gavin DG, Fisher DM, Herring EM, White A, Brubaker LB (2013) *Paleoenvironmental Change on the Olympic Peninsula, Washington: Forests and Climate from the Last Glaciation to the Present. Final Report to Olympic National Park.* University of Washington, Seattle, Washington. p. 109.

GSA (2018) *GSA Geologic Time Scale.* The Geological Society of America. Retrieved from: https://www.geosociety.org/documents/gsa/timescale/timescl.pdf. June 9, 2021.

Hunt CB (1988) *Geology of the Henry Mountains, Utah, as Recorded in the Notebooks of GK Gilbert, 1875–76.* 167. Geological Society of America, Boulder, Colorado.

Johnson DL, Johnson DN (2012) The polygenetic origin of prairie mounds in northeastern California. In:Horwath JL, Johnson DL (editors). *Mima Mounds: The Case for Polygenesis and Bioturbation, GSA Special Papers Volume 490.* The Geological Society of America, Boulder, Colorado.

Kahle SC, Olsen TD (1995) *Hydrogeology and Quality of Ground Water on Guemes Island, Skagit County, Washington.* US Geological Survey, Tacoma, Washington.

Keddy P (2010). *Wetland Ecology: Principles and Conservation* (2nd ed.). Cambridge University Press, Cambridge, United Kingdom.

Kruckeberg AR (1991) *The Natural History of Puget Sound Country.* University of Washington Press, Seattle, WA. pp. 61–64.

NSF (2003) *Rainfall Controls Cascade Mountains' Erosion and Bedrock Uplift Patterns.* ScienceDaily. Retrieved from: www.sciencedaily.com/releases/2003/12/031212075 857.htm. June 11, 2021.

O'Neil T (1949). *The Muskrat in the Louisiana Coastal Marshes.* Louisiana Department of Wildlife and Fisheries, New Orleans, Louisiana.

Reidel JL (2017) Deglaciation of the North Cascade Range, Washington and British Columbia, from the Last Glacial Maximum to the Holocene. *Cuadernos de Investigacion Geografica* 43(2): 467–496.

Reiners PW, Ehlers TA, Garver JI, Mitchell SG, Montgomery DR, Vance JS, Nicolescu S (2002) Late Miocene Exhumation and Uplift of the Washington Cascade Range. *Geology* 30(9): 767–770.

Reiners PW, Ehlers TA, Mitchell SG, Montgomery DR (2003) Coupled spatial variations in precipitation and long-term erosion rates across the Washington Cascades. *Nature* 426(6967): 645–647.

Rigg GB (1958) *Peat Resources of Washington, Bulletin No. 44.* Division of Mines and Geology, Department of Conservation, State of Washington, Olympia, Washington.

Seliskar DM, Gallagher JL (1983) *The Ecology of Tidal Marshes of the Pacific Northwest Coast: A Community Profile.* US Fish and Wildlife Service, Division of Biological Services, Washington DC.

Sherrod BL, Brocher TM, Weaver CS, Bucknam RC, Blakely RJ, Kelsey HM, Nelson AR, Haugerud RA (2004) Holocene fault scarps near Tacoma, Washington, USA. *Geology* 32: 9–12.

Shipman H (2008) *A Geomorphic Classification of Puget Sound Nearshore Landforms.* US Army Corps of Engineers, Seattle District, Seattle, Washington.

Shipman H (2010) The geomorphic setting of Puget Sound: Implications for shoreline erosion and the impacts of erosion control structures. In: Shipman H, Dethier MN, Gelfenbaum G, Fresh KI, Dinicola RS (editors). *Puget Sound Shorelines and the Impacts of Armoring – Proceedings of a State of the Science Workshop.* US Geological Survey, Washington DC.

Sloss CR, Hesp P, Shepherd M (2012) Coastal dunes: Aeolian transport. *Nature Education Knowledge* 3(10): 21.

Tabor RW, Cady WM (1978a) *Geologic Map of the Olympic Peninsula: US Geological Survey Map I-994.* US Geological Survey, Washington DC.

Tabor RW, Cady WM (1978b) *The Structure of the Olympic Mountains, Washington – Analysis of a subduction zone: US Geological Survey Professional Paper 1033.* US Geological Survey, Washington DC. p. 38.

USGS (1999) *Pilot Project: Mount Rainier Volcano Lahar Warning System.* US Geological Survey, Volcano Hazards Program, United States Geological Survey and Pierce County, Washington, Department of Emergency Management, Tacoma, Washington. Retrieved from: http://volcanoes.usgs.gov/About/Highlights/ RainierPilot/Pilot_highlight.html. June 10, 2021.

USGS (2014) *Significant Lahars at Mount Rainier.* United States Geological Survey, Washington DC. Retrieved from: https://www.usgs.gov/volcanoes/mount-rainier/ significant-lahars-mount-rainier. June 10, 2021.

van der Valk AG (1989). *Northern Prairie Wetlands.* Iowa State University Press. Ames, Iowa.

Webb NP, Galloza MS, Zobek TM, Herrick JE (2016) Threshold wind velocity dynamics as a driver of Aeolian sediment mass flux. *Aeolian Research* 20: 45–58.

WHSRN (2008a) *WHSRN Sites*. Western Hemisphere Shorebird Reserve Network, Manomet, Massachusetts.

WHSRN (2008b) *WHSRN: A Strategy for Saving Shorebirds*. Western Hemisphere Shorebird Reserve Network, Manomet, Massachusetts.

WSDNR (2021a) *Puget Lowland*. Washington State Department of Natural Resources, Olympia, Washington. Retrieved from: www.dnr.was.gov/program-and-services/geology/explore-popular-geology/geologic-provinces-washington/puget-lowland#glacial-features/. July 13, 2021.

WSDNR (2021b) *Puget Lowland*. Washington State Department of Natural Resources, Olympia, Washington. Retrieved from: www.dnr.was.gov/program-and-services/geology/explore-popular-geology/geologic-provinces-washington/puget-lowland#mima-mounds/. July 13, 2021.

5 Early Biology of Puget Sound

This chapter describes the changes in the flora and fauna of the Puget Sound region since the dawn of the region we call the Seattle-Tacoma Metropolitan Area. The majority of the underlying rocks of the Puget Sound region have been deposited in relatively recent times, in terms of the age of the Earth. Apart from some probable Paleozoic metamorphic rocks, the oldest rocks are marine sediments that date from the late-Triassic period, about 235 million years ago (mya) and mainly contain the remains of plankton, ammonites and belemnites, mollusks, and sea urchin with some terrestrial forms swept out to sea from the eastern landform. Due to tectonic uplift, the shallow seas gave way to terra firma by the beginning of the Miocene (about 10 mya) and land-dwelling mammals, reptiles, birds, and amphibians become more common. Many of these would appear unfamiliar to our eyes and it was not until the beginning of the Pliocene (5.33 mya) that the large mammals we recognize as being related to our contemporaries, namely mammoths, saber-tooth tigers, lions, hyenas, wolves, camels, horses, and deer, for example, appear in the fossil record. We end the chapter by summarizing the Great Extinction event that occurred towards the end of the last ice age (about 12,000 years ago) and the conditions that may have let this come about.

EVOLUTION OF THE PUGET SOUND REGION

In the previous chapters, we described the geology and geomorphology of the Puget Sound region during the past 50–100 million years. This chapter is devoted to the flora and fauna that existed during those times up until the beginning of the Holocene Epoch (11,700 years ago) and that were partly influenced by the underlying rock makeup, seascapes and landscapes that resulted from interactions between the underlying rock and minerals and the weathering elements. Those elements include wind, atmospheric vapor concentration, water and snow precipitation, followed by formation of streams and rivers, lagoons, wetlands, mesas, and valleys. We have chosen to include the Linnaean binomial (Genus species) nomenclature wherever we are able, so that the reader may explore elsewhere for more details if they find a topic of interest.

MESOZOIC ERA

Because parts of the Pacific Northwest as we know it did not exist before the dinosaurs went extinct, very few dinosaur fossils or any other animals of this Era

DOI: 10.1201/9780429487439-5

are found there, other than some marine mollusks, including cephalopods and bivalves, sharks, and marine reptiles, including plesiosaur remains (Murray, 1974; Ludvigsen and Beard, 1997). The geological history of the Puget Sound region was predominantly related to volcanic activity, accretion by terranes, whose origins upon the Farallon Plate suggest marine sediments and basaltic eruptions, as well as involving the Laramide orogeny.

All of these tectonic processes happened during the mid-to-late Triassic, the Jurassic, and early Cretaceous Periods, when the Farallon Plate was migrating adjacent to the western coast of Laurentia. Towards the middle of the Cretaceous, the Farallon Plate collided with Laurentia causing extensive tectonic uplift and deformation; the mixture of basaltic deposits, breccias, and marine sediments were accreted into their current location.

The oldest rocks in the Puget Sound region date from the early Triassic Period (252 to 247 mya) and are found in the San Juan Islands; they are chert and limestone sediments from the Triassic and Jurassic Periods. Examples are to be found on Orcas Island and San Juan Island and the limestone lenses are generally embedded within cherty formations. They probably originated as extremely fine particles of continental-derived eroded rock which sedimented far from shore as semi-colloidal silt as seen in the Orcas Group (248 to 150 mya) (Brandon et al., 1988). At the time the floor of the sea was mainly inhospitable to life although some corals (for example, *Lithostrotion*), conodonts, and radiolaria are found in these Triassic to mid-Jurassic-age sediments (McLellan, 1927).

The majority of the deposits from the late-Triassic and Jurassic Periods are either volcanic in origin or comprise mixtures of sedimentary conglomerates, breccias, grits, and slates, and metamorphic greywacke, argillite, schist, phyllite, and coal; they are found on San Juan, Orcas, and Lummi Islands (Leech River Complex, 100 to 88 mya) (Groome et al., 2003; Jakob and Joohnson, 2016). The foraminifer *Fusulina* is most abundant in these sediments (McLellan, 1927).

In the Mesozoic Era (252 to 66 mya) much of the emerging Pacific Northwest was covered with shallow inland seas in which lived a number of marine in-vertebrates and, most likely, marine reptiles. Sedimentary rocks from the upper Triassic Period (237 to 201 mya) are found, for example, in San Juan Island; they are primarily conglomerates, shale, slate, sandstone limestone, and metamorphic rocks. The San Juan Islands (and also Vancouver Island) result from accreted terranes derived from island-arc masses created close to the subduction boundary during the late Jurassic-early Cretaceous, followed by late Cretaceous thrusting (Brandon et al., 1988). The partially metamorphosed sedimentary rocks contain abundant impressions of fossils, such as *Halobia*, a bivalve (McLellan, 1927). There are some isolated island arc-derived formations from the mid- to late-Jurassic Period (174 to 145 mya) in the North Cascades and Fidalgo Island (Fidalgo Complex, 167 mya), but mainly comprise intrusive and metamorphic rocks (Livingston, 1959; MacDonald and Schoonmaker, 2017). During the Cretaceous Period (145 to 66 mya) foraminifers, coccoliths, ammonites, mol-lusks, and oysters, were present in the shallow seas (Kennedy and Moore, 1971).

Until 2012, no terrestrial vertebrates from the Mesozoic Era had been found in the Puget Sound region, but in that year, a team of paleontologists from the Burke Museum (University of Washington), searching for ammonites in the marine sandstones of Sucia Island, discovered a small portion of the left femur from a theropod dinosaur; it was subsequently named *Suciasaurus rex*, a tyranosaurid. It dates from the Late Cretaceous Period (80 mya) and is being nominated to be Washington's official state dinosaur (Peecock and Sidor, 2015; WSL, 2019).

As we observed earlier, for much of its geological history the Pacific Northwest was submerged beneath shallow coastal seas. From its creation during the late Jurassic Period (about 165 mya) until more recently (30 mya), the prehistoric western Washington region was largely affected by the subterranean motion of the Farallon Plate beneath that of the North American Plate, and which resulted in periodic volcanic activity as well as changes in the sea level caused by tectonic upward and downward motion of the continental plates. The entire Insular Superterrane (corresponding to present-day Vancouver Island, the Haida Gwai archipelago, and the Pacific coastline of British Columbia and Alaska) had accreted to the North American Plate by the early Paleogene (<66.5 mya). The Insular Terranes are believed to have begun as island arcs on the Farallon Plate, presumably having originally been formed above a subduction zone during the Carboniferous, the late Triassic, and the early Jurassic Periods (Johnston, 2013).

To the east of the Puget Sound region lie the Intermontane and the Coast Belts, also derived from earlier island arcs accreting on western Pangea during the early Jurassic and mid-Cretaceous Periods (Burke Museum 2021a, 2021b); the Intermontane Belt now underlies the western Rockies and the eastern Cascades whereas the Coast Belt underlies the San Juan Islands, the eastern Cascades, and Puget Sound. These too mainly comprise sedimentary and igneous rocks (e.g., plutons), interspersed by volcanic activity from the subduction of another eastward or northward moving oceanic plate under the main continental mass (Burke Museum 2021a, 2021b).

Much of the sedimentary rock laid down within the region date from the mid- to late-Mesozoic Era (165 to 66.5 mya) and for much of the Cenozoic Era (66.5 mya to the present day) and are embedded with mainly invertebrate fossils. These include mollusks such as belemnites (squid-like cephalopods), ammonites (nautilus-like cephalopods), bivalves (oysters, clams, and pelycopods, all having bilaterally symmetrical shells, i.e., their shells are upon their sides), and gastropods (including freshwater snails); crustaceans, such as crabs (decapods; Bishop, 1988; Clites, 2020) and acorn barnacles (*Sessilia* spp.); brachiopods (lamp shells, filter-feeders having dorsal/ventral shells that are distantly related to mollusks and annelid worms; Cohen and Weydmann, 2005); corals; sea urchins (such as sand dollars, order Clypeasteroida); and plankton, usually the larval forms of many invertebrate and some vertebrate species, and planktonic radiolaria (protozoa having mineralized skeletons) (Clites, 2020). What is not so readily apparent to the casual observer is the presence of microfossils, again usually larval forms of marine organisms; although invisible to the naked eye, they are of extreme importance to the energy industry geologist, for the presence (or absence) of a particular genus can indicate the co-localization of fossil fuels, such as oil and gas.

These marine organisms fluctuated both in numbers as well as across species, with the belemnites and ammonites becoming extinct at the end of the Cretaceous Period (66.5 mya), and increase in the numbers of crustacean species (Armstrong et al., 2009, Nyborg et al., 2003).

CENOZOIC ERA

While the sedimentary deposits from the Mesozoic Era are predominantly de-rived from volcanic submarine and terrestrial eruptions, and from marine or otherwise fluvial (rivers and lakes) conditions, the sedimentary rocks from the Cenozoic Era (from the mass extinction at 66.5 mya to the present day) comprise both aquatic and terrestrial species.

EOCENE EPOCH: 56 TO 33.9 MYA

The majority of the bedrock underlying the Puget Sound region is part of the Crescent Terranes (see Chapter 2) which began to accrete onto the side of the nascent North American Plate about 50 mya. Instead of accreting along a subduction zone, they were part of a transform tectonic process in which the colliding plate was moving in a northerly direction along the west margins of the North American Plate (Sharpe and Clague, 2006; Dawes and Dawes, 2013a). The rock and mineral makeup of this terrane strongly suggests that the formation derived from mantle-derived magma erupted from a spreading ridge and asso-ciated hotspot (as in today's Iceland) followed by magmas derived from sub-duction zone(s) (Phillips et al., 1989). Sediments from the Crescent Terrane at the base of the eastern Olympic Peninsula are of continental origin, suggesting that they derived from outflows off the North American Plate to the east.

Around 50 mya, the motion of the Pacific Plate appears to have undergone a momentous transition, the plate motion changing from a northerly directed motion to a northwesterly motion, as seen today (Dawes and Dawes, 2013a). At the same time, this would have had serious repercussions on the forces propel-ling the Farallon Plate, thereby changing the whole tectonic processes in the Pacific Northwest.

This appears to correlate with the change in the creation of volcanic island chains over a Pacific hotspot, now under the Big Island of Hawai'i (Hawaiian hotspot). At the most western end of the Hawaiian ridge, lies Midway Atol (*Pihemanu Kauihelani* in Hawaiian), the last remaining inhabitable island of the chain; all of the islands created earlier are now seamounts and form the Emperor seamount chain. The chain, created by the Hawaiian hotspot between 85 and 39 mya, form a line of seamounts crossing the entire northwestern Pacific seabed, eventually subducting under the Kuril-Kamchtka Trench, eastern Russia. Of note, around 50 mya, shown using geochronological dating, there was an abrupt change in apparent direction of the movement of the Pacific Plate over the hotspot, transitioning from a northerly motion to a northwesterly motion, in apparent synchrony with the movement of the Pacific and Farallon Plates in the Pacific Northwest. Two explanations have been under serious consideration since 1963: (i) that there was a real change in the direction of movement of the Pacific Plate and adjacent plate systems, or (ii) that the hotspot itself changed direction, due to fluctuations and deep mantle processes not yet understood (Wilson, 1963; Sharpe and Clague, 2006; Roach, 2003). Recently, the relative motion between three Pacific Ocean hotspots coupled with paleomagnetic data suggest that the Hawaiian hotspot has "moved" at different velocities during the past 60 million years and that a combination of the two explanations may be part of the solution (Konrad et al., 2018). The reader is invited to study this further.

The Olympic Peninsula results from another terrane accretion event, this time dating from between 55 to 15 mya; the terrane comprises a core of basaltic erupted lavas from underwater volcanos, whose flanks were overlain by sedimentary rocks, mostly of marine sandstone and siltstone (with occasionally conglomerate) from the mid-Eocene through the mid-Miocene (45 to 19 mya) (Tabor and Cady, 1978). About 10 mya, uplift formed the Olympic Mountains (Nesbitt and Scotchmoor, 2010).

The Cascades have been volcanic for over 36 million years, and thus very little evidence of prehistoric terrestrial life is found in the Puget Sound vicinity (Burke Museum, 2021b).

In general, the earliest sedimentary rocks found in the Puget Sound region from the Cenozoic are from the Eocene which was characterized by a warm wet climate resulting in warm shallow seas where we find marine and riverine fossil deposits, as well as those from terrestrial vertebrates (Walsh, 1996). Towards the end of the Eocene, the climate began to cool and the north and south poles began to accumulate ice (Peterson and Abbott, 1979, Frederiksen, 1991).

The Chuckanut Formation is a fluvial sandstone deposit dating from around 54 to 45 mya located on the mainland south of Bellingham, and is also exposed on Sucia Island in the San Juan archipelago (Brandon et al., 1988). The fossil flora and fauna remains reveal that the formation derived from rivers flowing through a subtropical rain forest, and include palms, giant ferns, sycamore, alder, insects, crocodiles, shore birds, large swamp-dwelling mammals (such as Pantodonta [extinct large-bodied hippo-like herbivores) and Dinocerata (brontotheres, extinct relatives of the rhino]), amphibians, and freshwater mussels. Between 45 and 25 mya, it underwent extensive folding resulting in at least four syncline/anticline pairs ; coal beds are also present (Mustoe and Gannaway, 1997). Of note, titanothere skeletons have been found in the Clarno Formation of Central Oregon, suggesting that these also inhabited the proto-Puget Sound region (Mustoe and Gannaway, 1997).

The Puget Group comprises the Renton Formation, comprising nonmarine sandstone and siltstone, shale, and coal, and the older Tukwila Formation that comprises volcanic sandstone, siltstone, shale, breccia, tuff, lahar, and carbonaceous shales, dating to about 42 mya; fossils of sharks, brachiopods, polychaetes, bivalves, crustaceans crabs, and sea urchins, are found there, suggesting a marine environment (Booth et al., 2021).

The Blakeley Formation of the late Eocene to mid-Oligocene (37 to 26 mya) comprises marine sandstones, shales, and conglomerates and is exposed in the sea cliffs along the entrance to Bremerton Inlet of Puget Sound and on Bainbridge Island. Fossil mollusks are typical animals found—there are also tree branches and leaves suggesting that the region was close to the shore. It is overlain by nonmarine conglomerates interbedded with sandstones and shales (Weaver, 1912, 1937; Fulmer, 1954; Johnson et al., 1994).

The Cowlitz Formation is from the mid- to late-Eocene (40 to 37 mya) and comprise sandstones, shales, conglomerates, and subordinate amount of lime-stone/shale, together with numerous intercalated layers of tuff and basaltic lava. They are partly marine and partly brackish water deposits. Fossils have been identified as pelecypods (clams and oysters), gastropods (snails and limpets), brachyopods, and some sharks' teeth (Weaver, 1912). Some aspects also cor-relate this formation with the Tejon Formation found in southern California (Henrikson, 1956).

The Lincoln Creek Formation encompasses the late Eocene through the Oligocene (38 to 16 mya) a period of more than 20 million years, an almost unbelievably long period of geological stability. It is a part of the Crescent Terranes and outcrops are found in the southeastern foothills of the Olympic Peninsula. The formation originates from tropical marine sediments of sandstone and tuffaceous siltstone having three foraminiferal zones: the *Sigmomorphina schencki* zone, the *Cassidulina galvinensis* zone, and the *Psuedoglandulina* aff. *P. inflata* zone (Beikman et al., 1967; Prothero and Armentrout, 1985). The formation contains many species of crustaceans and mollusks, often encased in a concretion (a ball of sedimentary rock which can be between 4 and 11 cm in diameter), as well as palms, conifers, gingko, banana, magnolia, species in the rose family, and grasses, suggesting that the environment was at times a brackish swamp close to a volcanic source (Sayce, 2012).

OLIGOCENE EPOCH: 33.9 TO 23 MYA

There are few sediments from the late Oligocene in the Puget Sound region, but some fossils have been found in the Carmanah Group in western Vancouver Island. These include bones and teeth from a primitive sea cow (*Cornwallius sookensis*, 25 mya), a bird limb bone, sharks' teeth, and bivalves in a breccia and conglomerates (Bream, 1987; Johnston, 2013).

As mentioned above, the upper portions of the Lincoln Creek Formation also date to this epoch; as we noted in Chapter 2, the Crescent Terranes are believed to have originated as a spreading margin, akin to the Iceland Plateau, although having a tropical climate. As an aside, contemporary Iceland has its origins during the middle Miocene (16 mya), an age comparable to the extensive period attributed to the Lincoln Creek Formation, however the offshore plateau would be expected to yield even older dates (Moorbath et al., 1968; Bott et al., 1983).

A fossil albatross, *Diomedavus knapptonensis*, has been recently reported from the Lincoln Creek Formation and is the oldest published albatross from the North Pacific Basin (Mayr and Goedert, 2017).

MIOCENE EPOCH: 23 TO 5.3 MYA

At the entry of the Miocene the western margins of the Pacific Plate were accumulating marine sediments, and we find a number of marine and coastal sedimentary formations within the Puget Sound region. The area that now comprises the Seattle-Tacoma metropolitan area and Puget Sound was mostly covered by warm seas, in which were found barnacles and other shellfish, shallow-water inhabitants. By the end of this time (about five mya), the Farallon Plate (by this time essentially the residual Juan de Fuca, Explorer and Gorda Plates) was almost entirely subducted beneath the North American Plate and the Pacific Plate had already accumulated marine sediments of its own and had become uplifted.

Farther to the south and east, there were sustained, vigorous basaltic eruptions from between 17 and 14 mya, resulting in the largest and most recent igneous province called the Columbia River Basalt Group; it covers over 210,000 km^2 and about half of the state of Washington (Dawes and Dawes, 2013a). Eruptions continued periodically for the next nine million years or so; there is evidence that the volcanic hotspot is now the source of the Yellowstone hotspot in Wyoming (Westby, ND). Note that these eruptions were distinct from the Cascade Range volcanic activity that had been in effect for around 36 million years.

The Astoria Formation of Washington comprises middle Miocene (23 to 16 mya) marine sediments; mollusks are a commonly found fossil in these deposits, as well as Nautilidae (cephalopod); in addition, charred petrified wood, leaves, pine cones, and a nut have also been identified (Moore, 1963; Smith and Manchester, 2018). Others have reported an early true seal, *Desmatophoca oregonensis*, from the Astoria Formation of Oregon (Deméré and Berta, 2002); the reason why the Washington formation of the same name has the "query mark" is that although it technically is not believed to be the same contiguous formation with its Oregon counterpart, it nevertheless has too many similarities to be derived from a different environment. Support for the two formations to be contemporaneous comes from the many identical mollusk species found in both the Washington and the Oregon formation (Moore, 1963). Recently, partial albatross fossil remains have been identified in the Washington formation ("Astoria Formation albatross"); however, the bones unearthed do not correspond to those found in the Lincoln Creek Formation of the Oligocene and thus has not yet been systematically named. Both are believed to be stem albatrosses that differ from extant species in the leg bones (Mayr and Goedert, 2017).

The Clallam Formation along the Juan de Fuca Strait is from lower Miocene (23 to 16 mya) nearshore sedimentary rocks (marine sandstone, sandy siltstone, granules, and conglomerates) comprising mollusks and foraminifera. Thin coal beds are present in the upper layers indicating non-marine deposition towards then of this succession; the molluscan fauna appears to correlate with the Astoria Formation species (Addicott, 1976).

The Montesano Formation of the upper Miocene (11.5 to 9.2 mya) marine sedimentary of siltstone, tuffaceous sandstone, conglomerates, and mudstone (USGS, ND). The Hammer Bluff Formation in the eastern Puget Lowland dates from between 16.5 and 5 mya having continental sedimentary rocks (both sandstone and volcanic ash) that include fossil leaves (Mullineaux, 1970).

An unnamed mid-Miocene formation (11.4 mya) that overlies the Eocene Blakeley Formation comprises mainly sandstone and conglomerates; terrestrial plant fossils are found in it, suggesting it was a fluvial deposit (Booth et al., 2021).

As we mentioned earlier, animals that inhabited the Miocene environments in the Puget Sound region are predominantly marine and riverine or estuary invertebrates, and these environments disappear from the fossil record as we enter the Pliocene Epoch (5.33 to 2.58 mya). This change in environment correlates with the uplift of the Puget Sound region resulting from the relatively northward-moving Pacific Plate slipping along the San Andreas Fault up the western edge of the North American Plate. Large shallow lakes formed in local structural basins during the late Miocene and sedimentation continued of lacustrine clay (settling particles from still water), fluvial sand, and volcanic mudflows (DNR, 1978; WSDNR, 2021a).

West of Puget Sound and the Olympic Mountains is the Grays Harbor region where fossilized remains of a Miocene seal (*Allodesmus demerei*; 10.5 to 9 mya) have been found; they measured about 3 m in length and probably weighed about 260 kg. The authors suggested that competition from early walruses may have contributed to the extinction of these early true seals (Boessenecker and Churchill, 2018).

Marked regional uplift of the Willapa Hills and deformation of the Olympic Mountains further reduced the area of marine deposition; uplift combined with fluvial outflow from the Olympics large areas of sediment were deposited on their western flanks. Terrestrial sediments were thus forming on the eastern margins of the Pacific Plate and the western margins of the North American Plate and we find a plethora of plants and animal fossils from Pliocene sediments.

It is also somewhat likely that the ancestors of Pacific salmon were present in the streams, rivers, and waterways during this time. At least nine lineages of salmon (*Onchyrynchus*, *Prosopium*, and *Paleolox* spp.) have been reported from the western Snake River (southeastern Oregon) and the Great Basin in Mio-Pleistocene sediments, including a small lake-dwelling relative of the sabertooth (or spiketooth) salmon (O. ["*Smilodonichthys*"] rastrosus; 8.7 to 5.3 mya) of the Columbia River drainage area (Stearley and Smith 2016).

PLIOCENE EPOCH: 5.3 TO 2.6 MYA

Much of the uplift of the Cascade Range took place during the Pliocene; the most recent volcanic episodes began about 5 mya (USGS, 2004). Much of the tectonic activity in the Pacific Northwest occurred east of the Cascades in the Basin and the Rocky Mountains, the basaltic eruptions of the Columbia River Basalt Group. The only major source of volcanic activity close to Puget Sound was in the Great Rocks area, southeast of Mount Rainier (Dawes and Dawes, 2013a).

Continued uplift of the region further moved the shoreline westwards during the early Pliocene (DNR, 1978). Shallow-water marine echinoderms and mollusks from this period are found in the Puget Sound region (Woodring, 1938; Addicot, 1973). Deposition continued on the west coast of the Olympic Peninsula during the Pliocene and uplift created new terrestrial habitats (WSDNR, 2021b).

The climate was also changing; the conditions varied from very warm, equitable climates to cooler periods in cycles of approximately 40,000 years; atmospheric carbon dioxide levels reached 425 ppm in the early Pliocene but then temperature and humidity decreased gradually until the extreme Ice Age cycles began about two million years ago at the beginning of the Pleistocene (Burke et al., 2018).

A review of the literature suggests that there was a distinct lack of terrestrial fauna during the Pliocene in the Puget Sound region. This may be due to a number of factors, including, but not limited to, ongoing Cascade volcanics and the continued eruptions from the Columbia River Basalts, resulting in a less-than-ideal habitable ecosystem.

In order to address the likely flora and fauna, we will use published records from eastern Washington and northwestern Oregon to give a sense of what may have been in the vicinity (Gustafson, 1978; Martin, 2008).

Typical early Pliocene animal fossils include the traditional megafauna of North America. The term "megafauna" is considered to include megaherbivores (>1000 kg) and megacarnivores (>100 kg) (Malhi et al., 2016). In the early Pleistocene, these megaherbivores included camels (*Camelops*), ground sloths (*Megalonyx*), glyptodonts (*Glyptotherium*), toxodonts (*Toxodon*), oreodonts (Fam. Merycoidodontidea), and mastodons (*Mammut pacificus*), all of which evolved in the Americas. Large herbivores (45–999 kg) included primitive horses (such as the three-toed *Hipparion forcei* and *Merychippus californicus*), oreodonts, a medium-sized even-toed ungulate possibly related to camels, all now extinct (Spaulding et al., 2009); large carnivores (21.5–99 kg) included saber-tooth big cats (such as *Smilodon*), American lions (*Panthera atrox*), canids (*Canis lepophagus* and *C. edwardii*, possibly the ancestors of the coyote and wolf, respectively), and hyenas (*Chasmaporthetes ossifragus*). Smaller mammals included foxes (*Vulpini*), giant beavers (*Castoroides nebrascensis*), primitive ground squirrels, mustelids (e.g., badgers, martens, and otters), peccaries (*Platygonus* and *Mylohyus*), raccoon-like animals (*Procyon* spp.), and birds and lizards of various species. The middle Pliocene fauna comprised beardogs, hyenas, camels, flamingos (*Phoenicopterus copei* and *P. minutus*), ground sloths, mastodons, pronghorn antelope (*Antilocapra americana*), rhinoceroses (*Teleoceras*), cougar-like cats (*Puma pumoides*), and small rodents. The late Pliocene saw the appearance of many of North America's modern animal assemblage including bison (*Bison bison*), horses (*Equus* spp.), elk (*Cervus* spp.), and moose (*Alces americanus*) (Murray, 1974).

Most of the environmental niches of today were filled, but with different animals. The terrestrial environment was dominated by mammals and birds. The largest were the elephant and mastodon relatives called *Gomphotherium*, and they lived alongside dozens of other large mammals, such as three-toed horse *Hipparion* the tiny pronghorn *Merycodus*, the long-necked camel *Aepycamelus* and peccaries. There were also smaller mammals, such as rabbits, beavers, ground squirrels, foxes, and moles. Birds and reptiles shared the space. The large number of herbivores supported a significant number of predators. The dog *Borophagus* probably killed and also ate carrion. *Nimravides* was a predatory cat. The largest predator was the sabertooth cat *Barbourofelis*.

Great American Biotic Interchange

The Great American Biotic Interchange (GABI) was an important paleozoo-geographic event in which animals and plants were exchanged between the North and South American continents around the end of the Pliocene Epoch. Three million years ago, tectonic forces and subduction of the Cocos Plate under the Caribbean and North American Plates formed the Panama isthmus between the Americas (O'Dea et al., 2016).

The GABI thus represented a north-south movement of eutheria from North America to South America [cats, camels (e.g., vicuñas, the parent species of llamas), tapirs, and peccaries] and northward flow from South America of opossums and armadillos (O'Dea et al., 2016).

The Pleistocene (beginning about 2.6 mya) ushered in a long period of climatic instability, including successive cycles of glacial (ice ages) and interglacial (warmer periods) periods (Cohen et al., 2013). The reasons for this climatic instability are complex, but have been associated with wobble and periodic reversals of the Earth's magnetic field; fluctuations in the amount of energy from the sun, most likely due to eccentricities in the Earth's orbit around the sun; and tectonic or volcanic activity (Buis, 2020; Nordt et al., 2003; Foulger, 2010).

The presence of marsupials in the Americas may be seen to some as odd, since marsupials are usually associated as Australasian fauna. This is where an understanding of palaeontology and geology bring to light the most likely explanation. During the early Cretaceous period, around 126 mya, the landmasses of North America, Europe, and Asia (composing the supercontinent of Laurasia) were in the process of breaking up; South America and Australia were still connected by Antarctica as part of the larger supercontinent Gondwana. At that time dinosaurs were the dominant large animal group and the ancestors of the marsupials, termed metatherians, were generally small and most likely nocturnal. Unfortunately, there is scant fossil evidence for these marsupial ancestors living on all three southern continents at that time; only monotremes and their ancestors have been found (Benson et al., 2013). It is most likely that the metatherians evolved in Asia, subsequently moved into North America by way of Europe, and then, at the end of the Cretaceous (66 mya), crossed by a landbridge to South America and then onwards expanding into Gondwana (Flynn and Wyss, 1998; Flynn et al., 2007). Plate tectonics then caused the landbridge to migrate north-east, where it became part of the Caribbean Archipelago, thereby cutting off any further connections between the continents of Laurasia and Gondwana (Kemp, 2005; Boschman et al., 2014). There is, however, also evidence of non-placental therian fossils already in Gondwana during the mid-late Cretaceous (83.6 to 66 mya; Newham et al., 2014).

Pleistocene and Holocene Epochs: 2.6 MYA to the Present

The Quaternary Glaciation began about 2.58 mya and is ongoing (Marshall 2010). From then to the present, the Earth has experienced several glacial periods. In that time, there have been several cycles of expansion and retreat of ice. In the colder parts of the cycle, ice sheets and glaciers covered a significant portion of the Northern Hemisphere. Temperatures and sea levels fell. Between those glacial periods, temperatures and sea levels rose. For much of the Quaternary Period (2.58 mya to the present), the ice cycle repeated about every 41,000 years, but about a million years ago, the cycle changed to every 100,000 years. The reasons are not quite clear. In fact, the cause of the cycles themselves are not known, but seem to be related to the tilt of the Earth.

In the Puget Sound region, at least sixcycles ofglacial and interglacial periods have been recorded, but these may be a fraction of the total when the marine-isotope record is taken into consideration; for example there are at least three stade/interstades (short periods ofglacial advance/retreat, e.g., the Younger Dryas) between 450,000 and 280,000 years ago (Booth, Troost, Clague, Waitt 2003). The last time the Cordilleran Ice Sheet advanced and retreated (Fraser Glaciation, 25,000 to 10,500 years ago) it obliterated some of the features that earlierglacial periods had created, so much of what we know now is pertinent onlyto the past ten thousand years or so. During maximum glaciations in the past, glaciers extended from the Olympic Mountains to the coastal plain, so these have left us with some indication of the duration and effects upon the environment (Booth, Troost, Clague, Waitt 2003; Dawes and Dawes 2013c). The flora and fauna we now describe is probablyrepresentative of the life present in the Puget Sound region duringthe interglacial periods, but much of the fossil record may have been obliterated bythe cyclicalglacial advances.

The Pleistocene Epoch (from 2.6 million to about 11,000 years ago) includes the climatic changes that brought about the periodic ice ages, as we discussed in Chapter 2 and above. It is here that the mammoths finally enter the North American ecosystem, having evolved around five million years ago in Africa from a common ancestor of the Asian elephant (Lister and Bahn, 2007; Krause et al., 2006). It is considered that the Columbian mammoth (*Mammuthus columbi*) in North America is descended from hybrids of either the steppe mammoth (*Mammuthus trogontherii*) or the woolly mammoth (*Mammuthus primigenius*) with the Krestovka mammoth (*Mammuthus* unk. sp.) in Siberia and the resulting hybrid(s) then entered North America over one million years ago before finally evolving into the Columbian mammoth (Lister and Sher, 2015; van der Valk et al., 2021). The woolly mammoth (*M. primigenius*) also had evolved in Siberia and subsequently migrated across Beringia during an interglacial period possibly by around 400 kya and then also interbred with the Columbian mammoth (Lister and Sher, 2015; van der Valk et al., 2021). The Pleistocene Epoch also saw the appearance of the American lion (*Panthera atrox*), American cheetah (*Miracinonyx* spp.), giant jaguar (*Panthera onca augusta*), puma (*Puma concolor*), lynx (*Lynx rufa*), dire wolf (*Canis dirus*), timber or gray wolf (*Canis lupus*), gray fox (*Urocyon cinereoargenteus*), short-faced bear (*Arctodus simus*), modern horse (*Equus ferus/caballus*), steppe bison/bison (*Bison priscus/Bison bison*), reindeer/caribou (*Rangifer tarandus*), shrub-ox (*Euceratherium collinum*), musk ox (*Ovibos moschatus*), tapir, pronghorn, elk, mule deer, bighorn sheep, and peccaries (Lorenzen et al., 2011; Oberbauer, 2018; Chimento and Dondas, 2018).

During the late Pleistocene and the early Holocene, plants and animals evolved very little, but there were big changes in their populations and especially among the megafauna. The mammals included mammoths, mastodons, camels, horses, llamas, elk, tapirs, moose, and bison, along with large predators, such as the short-faced bear, saber-tooth cat, wolves and American lions.

All of these species and more died out and were not replaced by others as was the normal expectation. The reason is not clear. Some scientists suggest that climate change was responsible (Wroe and Field, 2006; Guthrie, 2003). However, others focus on the expansion of humans and their arrival in North America. Barnosky et al. (2016) weighed these two theories. First, they found that the average temperature changes and anomalous precipitation and the velocity of those changes 132,000 to 1000 years ago and found no correlation with the loss of mammals. From these results, they discount the effects of climate change on the disappearance of the megafauna in the early Holocene. Second, they evaluated the effects of humans on the animals. Their results suggest that early hominoids in Africa and those that left Africa (e.g., Neanderthals and Denisovans) co-evolved with the large mammals there. Once modern humans (*Homo sapiens*) began to leave Africa and spread into Australia, Northern Asia and Europe, and North and South America, they were the first humans to have contact with the mammals in those areas. Modern humans brought new tools and hunting techniques that easily and quickly overwhelmed the large mammals.

About 100,000 years ago, the Late Pleistocene Epoch began. At that time, large herds of mammoth, mastodons, camels, horses, bison, llamas, elk, and tapirs were present (Hoppe, 2004; Grayson and Meltzer, 2015; Oberbauer, 2018). Over them all soared the condor (*Terratornis merriam*) with its wingspan of over 3 m. The herds likely moved from inland in the winter to the shore in the summer to escape the heat and eat the more vegetation kept lush by the fog. These would have been fairly short migrations of less than 60 km.

Among the large mammals were the Columbian mammoth (*Mammuthus columbi*), which weighed more than 5 tonnes and stood 4 m at the shoulders. Long-horned bison (*Bison latifrons*) and ancient bison (*Bison antiquus*) both grazed on grasses. Both were larger than modern bison. The long-horned bison lived alone or in small groups, but the ancient bison were herd animals. The American mastodon (*Mammut americanum*) was distantly related to elephants; it was shorter and stockier than the mammoth. The western horse (*Equus occidentalis*) was small and stood only about 1.5 m at the shoulders. The giant horse (*Equus pacificus*) was larger. Interestingly, the large-headed llama (*Hemiauchenia macrocephala*) was a grazer and far from the Andes that are normally associated with llamas. Jefferson's ground sloth (*Megalonyx jeffersoni californicus*) was as large as an ox. Camel (*Camelops hesternus*) were herd animals of about the same size as modern camels.

Well-fed predators were there too, including the short-faced bear, saber-tooth cat, dire wolf, and American lion. The dire wolf (*Canis dirus*) was about the size of a modern-day timber wolf and probably hunted in packs as do modern wolves. The large number of dire wolves found in the La Brea Tar Pits suggests that they ate carrion as well as hunted. Recent studies of dire wolf DNA from sub-fossil remains have found that when compared with modern wolves, coyotes, and jackals, they are genetically quite distinct from that group, having split form the Eurasian canids about 5.7 mya (Perri et al., 2021). The authors suggested that this supports the taxonomic classification under the prior name *Aenocyon dirus*. The coyote, *Canis orcutti*, is now extinct. A large number of them were also found in the La Brea Tar Pits, and they were probably significant predators. They were larger than their modern counterparts, but could not make adjust after most of the larger mammals were eliminated. The giant short-faced bear (*Arctodus simus*) was the largest carnivore that ever lived in the New World. An adult weighed about 900 kg and was a third larger than a grizzly bear. In addition, its longer legs suggest that it was faster even than modern bears. The cat family was well represented. The scimitar cat (*Homotherium serum*) was closely related to the saber-tooth cat (*Smilodon californicus*) and hunted large animals. In addition to many of the modern cat forms, the saber-tooth cats were major predators. The American lion (*Panthera atrox*) was larger than modern-day cats in Eurasia and may have hunted in prides as modern lions. The jaguar (*Panthera onca*), American cheetah (*Acinonyx trumani*), cougar (*Felis daggetti*), and lynx (*Lynx rufa fischeri*) rounded out this group.

By the beginning of the ice age that began about two million years ago, many of the earlier mammals, including 80 genera, had disappeared. As is almost always the case, when one genus disappears, another takes its place. Those early mammals were replaced by mammoths, dire wolves, and sabertooth cats. Some sort of extinction event seems to have eliminated many of the living organisms, but not to the extent that it could be considered a mass extinction event. The cause is unknown. Some speculate that climate change was involved. At about that time, the Earth cooled, the ice sheets at the polar caps expanded, glaciers expanded southward, rainfall was reduced, and the oceans cooled and sea levels fell. Other complications might have added to the stress on the mammals. Climate change also affected the vegetation. Grasslands gave way to a different environment more like our current one that features chaparral, multiple grasses, and many new species. Perhaps some combination of all of these was responsible for the loss of the genera. Some have speculated that the arrival of Native

Americans might have contributed to the demise of many of these species. However, that hypothesis remains unproved at this point.

The last of these ice ages began 33,000 years ago and started to recede 19,000 years ago. At its glacial peak, the Puget Sound trough and the Juan de Fuca Strait were entirely covered by lobes from the Cordilleran Ice Sheet (DNR, 2021; Puget Lobe, Juan de Fuca Lobe). During that time, sea levels were much lower than before, and a land bridge was revealed between Siberia and Alaska. Native Americans crossed that bridge to become the first humans in the Americas. They made it to the Puget Sound region about 15,000 years ago.

The Great Megafauna Extinction

As the last ice age began to come to an end about 15,000 years ago, almost all the megafauna of North America became extinct. The causes of these extinctions have been debated amongst scientists for many decades and have revolved from (1) adaptations (or lack of) to climate change, (2) a reduction in habitat resulting in reduced food supply for these huge mammals, (3) the appearance of man, who may have hunted them to extinction, or (4) a combination of all these factors (Clutton-Brock, 1996; Barnosky et al., 2004; Martin, 1984; Alroy, 2001; Hoppe, 2004; Stuart et al., 2004; Koch and Barnosky, 2006; Sandom et al., 2014; Surovell et al., 2016). A study that modeled population dynamics of six mega-fauna species over the past 50,000 years suggested that climate change and population fragmentation, and for certain species, human engagement, played the major role in their extinction in North America (Lorenzen et al., 2011). The remaining medium-to-large herbivores, such as bison, elk, moose, reindeer, and pronghorn antelope, survived this period of change, for reasons as yet unknown; this may be a combination of surviving in large herds, adapting to the relatively warmer climates and the multiplicity of food that was available in the woods and grasslands that came to dominate the landscape after the ice age. Musk ox survived in the Canadian and Alaskan Arctic (Lorenzen et al., 2011). In addition, populations of moose (*Alces*) and elk (*Cervus* spp.) increased towards the end of the Late Quarternary Period (late Pleistocene) that may have led to increased competition for both woodland and grassland resources (Guthrie, 2006). Of note, both reindeer (in Eurasia) and caribou (North America) have been hunted and/or herded by Arctic peoples for at least the past 10,000 to 15,000 years (Kurtén, 1968) thus perhaps inadvertently preserving pan-regional breeding populations.

Many cases of animals going extinct or nearly extinct are well known. The Columbian mammoth (*Mammuthus columbi*) lived in the Puget Sound region during the Late Pleistocene. These relatives of modern elephants stood about 4 m tall, weighed 11,000 kg with tusks almost 5 m long. They migrated into North America 2 mya but became extinct about 11,000 years ago. Other mammals included 2-tonne bison, Western camels, giant horses, tapirs and giant ground sloths. There were probably many American lions that lived in what is now the Pacific Northwest. They weighed 340 kg and were much larger than modern African lions. The other major predator was the short-faced bear. They were 4 m long, weighed 900 kg and could probably run at 65 kph for just over 1 km. Other predators included dire wolves, saber-toothed cats, grizzly bears, and giant condors.

Washington and British Columbia originally had two types of bears. The black bear (*Ursus americanus*) may have descended from the Pleistocene *Ursus otimus*. The grizzly bear (*U. arctos horribilis*) was found throughout the region (Mustoe and Carlstad, 1995). The grizzlies were larger than the grizzlies that lived in the Rockies (900 kg as compared to 700 kg).

We will now present a brief description of the flora associated with the period from about 10 mya to the beginning of the Holocene Epoch, about 11,700 years ago.

FLORA

In the late Miocene and early Pliocene epochs (about seven to five million years ago), a change occurred in the ratio of plants using the C3 and C4 photosynthetic pathways. For most plants on Earth, the first product of photosynthesis is a three-carbon product. In this process, because CO_2 enters through the stomata, and when the stomata are open, the plant can lose significant amounts of water. During a drought, this is a disadvantage. Plants in hot, dry areas have evolved C4 photosynthesis, which results in a four-carbon product. This process allows plants to carry on photosynthesis with the stomata closed. C4 plants include maize, sugarcane, and sorghum. The C4 process is also favored under conditions of low carbon dioxide levels. Cerling et al. (1997) examined the teeth of grazing animals during that period and determined that a significant change in the diets of those animals indicate a change in the atmospheric CO_2 levels that favored C4 photosynthesis.

Plants were highly diverse by the time of the dinosaurs (Millar and Woolfenden, 2016). Conifers appeared about 200 mya. The pines were the last to develop, and they were common about 145 mya. Gymnosperms, including ferns and horsetails, expanded worldwide about 150 mya. Angiosperms, traced back to 110 to 150 mya, had expanded to become the most diverse group of plants on Earth. Angiosperms were in eastern Washington at 100 to 120 mya, but their diversity can be pegged to about 55 mya. However, temperatures and humidity rose from 50 to 52 mya, and the angiosperms bloomed (no pun intended). Species found in more tropical areas were common, including avodado, palm, viburnum, magnolia, jackfruit, and figs (Millar and Woolfenden, 2016). Pollen was found for pine, walnut, hickory, and sweetgum.

At around 33 mya, the tropical species disappeared and more warm temperature species again became dominant. Cypress, Douglas-fir, evergreen oaks, sycamore, cottonwood, willow, redbud, barberry, cherry, ironwood, manzanita, flannel bush, sumac, and grasses were probably present in southern Washington (Herman, 1985; Mustoe and Leopold, 2014).

Regarding plant species, at the boundary between the Miocene and the Pliocene (about 5.3 million years ago), alder, cherry, Christmas berry, chumico, coffee berry, Douglas-fir, dogwood, elm, flannel bush, lilac, magnolia, mountain mahogany, manzanita, live oak, poplar, bush poppy, swamp cypress, sumac, desert sweet, sycamore, tupelo, and willow all grew around the Puget Sound region (Murray 1974,Addicot 1973; Herman, 1985).

After the end of the Ice Age, the eastern slopes of the Olympic Mountains saw the appearance of Western hemlock (*Tsuga heterophylla*), some Pacific silver fir (*Abies amabilis*), and Douglas-fir (*Pseudotsuga menziensii*) creating a similar coniferous forest that we see today (Gavin et al., 2013).

REFERENCES

Addicot WO (1973) *Neogene Marine Mollusks of the Pacific Coast of North America: An Annotated Bibliography, 1797-1969.* Geological Survey Bulleting 1362, Geological Survey, Department of the Interior, Washington DC.

Addicott WO (1976) *Molluscan Paleontology of the Lower Miocene Clallam Formation, Northwestern Washington.* US Geological Survey, Department of the Interior, Washington DC, p. 44.Retrieved from: http://pubs.er.usgs.gov/pubs/pp/pp976.

Alroy JA (2001) Multispecies overkill simulation of the end-Pleistocene megafaunal mass extinction. *Science* 292: 1893–1896.

Armstrong A, Nyborg T, Bishop G, Osso A, Vega FJ (2009) Decapod crustaceans from the Paleocene of central Texas, USA. *Revista Mexicana de Ciencias Geologicas* 26(3): 745–763.

Barnosky AD, Koch PL, Feranec RS, Wing SL, Shabel AB (2004) Assessing the causes of late Pleistocene extinctions on the continents. *Science* 306: 70–75.

Barnosky AD, Lindsey EL, Willavicencio NA, Bostelmann E, Hadly EA, Wanket J, Marshall CR (2016) Variable impact of late-Quaternary megafaunal extinction in causing ecological state shifts in North and South America. *Proceedings of the National Academy of Sciences of the United States of America* 113: 856–861.

Beikman HM, Rau WW, Wagner HC (1967) *The Lincoln Creek Formation Grays Harbor Basin Southwestern Washington.* US Geological Survey, US Department of the Interior, Washington DC.

Benson RBJ, Mannion PD, Butler RJ, Upchurch P, Goswami A, Evans SE (2013) Cretaceous tetrapod fossil record sampling and faunal turnover: Implications for biogeography and the rise of modern clades. *Paleogeography, Paleoclimatology, Paleoecology* 372: 88–107.

Bishop GA (1988) Two crabs, *Xandaros sternbergi* (Rathbun 1926) n. gen., and *Icriocarcinus xestos* n. gen., n. sp., from the Late Cretaceous of San Diego County, California, USA, and Baja California Norte, Mexico. *Transactions of the San Diego Society of Natural History* 21: 245–257.

Boessenecker RW (2013) A new marine vertebrate assemblage from the Late Neogene Purisima Formation in Central California, part II: Pinnipeds and Cetaceans. *Geodiversitas* 35(4): 815–939.

Boessenecker RW, Churchill M (2018) The last of the desmatophocid seals: A new species of *Allodesmus* from the upper Miocene of Washington, USA, and a revision of the taxonomy of Desmatophocidae. *Zoological Journal of the Linnean Society* 184(1): 211–235.

Booth DB, Walsh TJ, Troost KG, Shimel SA (2021) *Geologic Map of the East Half of the Bellevue South 7.5' x 15' Quadrangle, Issaquah Area, King County, Washington.* U.S. Geological Survey, Washington DC. July 16, 2021.

Booth, DB, Troost, KG, Clague, JJ, & Waitt, RB (2003). The Cordilleran ice sheet. Developments in Quaternary Science, 1, 17–4310.1016/S1571-0866(03)01002-9.

Boschman LM, van Hinsbergen DJJ, Torsvik TH, Spakman W, Pindell JL (2014) Kinematic reconstruction of the Caribbean region since the Early Jurassic. *Earth-Science Reviews* 138: 102–136.

Bott MHP, Saxov S, Talwani M, Thiede J (1983). *Structure and Development of the Greenland-Scotland Ridge: New Methods and Concepts.* Springer Science +Business Media. Plenum Press, New York NY p. 464.

Brandon MT, Cowan DS, Vance JA (1988) *The Late Cretaceous San Juan Thrust System, San Juan Islands, Washington.* Special Paper 221, The Geological Society of America, Inc., Boulder, Colorado.

Bream SE, (1987) *Depositional Environment, Provenance, and Tectonic Setting of the Upper Oligocene Sooke Formation, Vancouver Island, B. C.* WWU Graduate School Collection, Western Washington University. p. 659.

Buis A (2020) Milankovitsch (Orbital) Cycles and Their Role in Earth's Climate. NASA. Retrieved from: https://climate.nasa.gov/news/2948/milankovitch-orbital-cycles-and-their-role-in-earths-climate/. September 28, 2020.

Burke KD, Williams JW, Chandler MA, Haywood AM, Lunt DJ, Otto-Bliesner BL (2018) Pliocene and Eocene Provide Best Analogs for Near-Future Climates. *Proceedings of the National Academy of Sciences USA* 115(52): 13288–13293.

Burke Museum (2021a) *The Omineca Episode (180-115 Million Years Ago).* Burke Museum of Natural History and Culture, University of Washington, Seattle, Washington. Retrieved from: www.burkemuseum.org/geo_history_wa/ TheOmineca Episode.htm. July 14, 2021.

Burke Museum (2021b) *The Cascade Episode, Burke Museum of Natural History and Culture.* University of Washington, Seattle, Washington. Retrieved from: www.burkemuseum.org/geo_history_wa/CascadeEpisode.htm. July 14, 2021.

Cerling TE, Harris JM, MacFadden BJ, Leakey MG, Quade J, Eisenmann V, Ehleringer JR (1997) Global vegetation change through the Miocene/Pliocene boundary. *Nature* 389: 153–158.

Chimento N, Dondas A (2018) First record of *Puma concolor* (Mammalia, Felidae) in the early-middle pleistocene of South America. *Journal of Mammalian Evolution* 25(Suppl. 1): 1–9. DOI: 10.1007/s10914-017-9385-x.

Clites E (2020) *Fossils in our Parklands.* University of California Museum of Paleontology. Retrieved from: https://ucmp.berkeley.edu/science/parks/golden_gate.php. December 9, 2020.

Clutton-Brock J (1996) Horses in history. In:Olsen S (editor). *Horses Through Time*, pp. 83–102. Roberts Rinehart Publishers, Dublin, Republic of Ireland.

Cohen KM, Finney SC, Gibbard PL, Fan J-X (2013) *International Chronostratigraphic Chart 2013.* Stratigraphy.org. January 7, 2019.

Cohen BL, Weydmann A (2005) Molecular evidence that phoronids are a subtaxon of brachiopods (Brachiopoda: Phoronata) and that genetic divergence of metazoan phyla began long before the early Cambrian. *Organisms Diversity & Evolution* 5: 253–273.

Dawes RL, Dawes CD (2013a) *Geology of the Pacific Northwest Lecture. Lecture 8 – The Cenozoic era in the Pacific Northwest.* Retrieved from: https://commons.wvc.edu/ rdawes/Lectures/lect8.html. July 17, 2021.

Dawes, RL, & Dawes, CD (2013b). *Geology of the Pacific Northwest Lecture. Lecture 8 - The Cenozoic era in the Pacific Northwest*, 17, 2021.

Dawes, RL, & Dawes, CD (2013c). *Geology of the Pacific Northwest Lecture. Lecture 2 - The Quaternary Period in the Pacific Northwest, Pleistocene ice ages to the present,* Retrieved from https://commons.wvc.edu/rdawes/Lectures/lect2.html. 17, 2021.

Deméré T, Berta A (2002) The Micoene pinniped Desmatophoca oregonensis Codon, 1906 (Mammalia: Carnivora), from the Astoria Formation, Oregon. In:Emry RJ (editor). *Cenozoic Mammals of Land and Sea: Tributes to the Career of Clayton E. Ray. Smithsonian Contributions to Paleobiology,* pp. 113–147. Smithsonian Institution Press, Washington DC.

DNR (1978) *Geology of Washington.* USGS and Washington Department of Conservation, Department of Natural Resources, Olympia, Washington. Retrieved from: https://dnr.wa.gov/publication/ger_reprint12_geol_of_wa.pdf. July 19, 2021.

DNR (2021) *Extent of the Puget Lobe of the Cordilleran Ice Sheet During the Latest Ice Advance.* Department of Natural Resources, Olympia, Washington. Retrieved from: www.dnr.wa.gov/picture/ger/ger-explor_puget_glaciation.png. July 19, 2021.

Flynn JJ, Wyss AR (1998) Recent advances in South American mammalian paleontology. *Trends in Ecology and Evolution* 13(11): 449–454. doi: 10.1016/S0169-5347(98) 01457-8. PMID 21238387.

Flynn JJ, Wyss AR, Charrier R (2007) South America's missing mammals. *Scientific American* 296(May): 68–75.

Foulger GR (2010) *Plates vs. Plumes: A Geological Controversy.* Wiley-Blackwell, Heboken, New Jersey.

Frederiksen NO (1991) Pulses of Middle Eocene to Earliest Oligocene climatic deterioration in Southern California and the Gulf Coast. *Palaios* 6(6): 564–571.

Fulmer CV (1954) Stratigraphy and paleontology of the type Blakeley formation of Washington [abs.]. *Geological Society of America Bulletin* 65(12–2): 1340–1341.

Gavin DG, Fisher DM, Herring EM, White A, Brubaker LB (2013) *Paleoenvironmental Change on the Olympic Peninsula, Washington: Forests and Climate from the Last Glaciation to the Present. Final Report to Olympic National Park.* University of Washington, Seattle, Washington. p. 109.

Grayson DK, Meltzer DJ (2015) Revisiting Paleoindian exploitation of extinct North American mammals. *Journal of Archaeological Science* 56: 177e193.

Groome WG, Thorkelson DJ, Friedman RM, Mortensen JK, Massey NWD, Marshal DD, Layer PW (2003) *Magmatic and Tectonic History of the Leech River Complex, Vancouver Island, British Columbia; Evidence for Ridge-Trench Intersection and Accretion of the Crescent Terrane.* The Geological Society of America, Inc., Boulder, Colorado.

Gustafson EP (1978) The vertebrate faunas of the Pliocene Ringold Formation, South-Central Washington. In: *Bulletin of the Museum of Natural History, University of Oregon,* no. 23, p. 72. Museum of Natural History, University of Oregon, Eugene, Oregon.

Guthrie RD (2003) Rapid body size decline in Alaskan Pleistocene horses before extinction. *Nature* 426: 169–171.

Guthrie RD (2006) New carbon dates link climatic change with human colonization and Pleistocene extinctions. *Nature* 441: 207–209.

Henrikson DA (1956) *Eocene Stratigraphy of the Lower Cowlitz River-Eastern Willalpa Hills Area, Southwestern Washington.* Department of Conservation and Development, State of Washington, Tacoma, Washington.

Herman RK (1985) *The Genus Pseudotsuga: Ancestral History and Distribution.* Forest Research Laboratory, Special Publication 2b, Oregon State University, Corvallis, Oregon.

Hoppe KA (2004) Late Pleistocene mammoth herd structure, migration patterns, and Clovis hunting strategies inferred from isotopic analyses of multiple death assemblages. *Paleobiology* 30: 129–145.

Jakob J, Joohnson ST (2016) *The Leech River Complex – A Cretaceous Accretionary Wedge*. Conference: Geological Association of Canada/Mineralogical Association of Canada (GAC/MAC) 2016 – Margins Through Time, White Horse, Yukon, Canada.

Johnson SY, Potter CJ, Armentrout JM (1994) Origin and evolution of the Seattle fault and Seattle basin, Washington. Geological Society of America. *Geology* 22(1): 71–74.

Johnston S (2013) *Fossils of Vancouver Island*. Poster, Earth Science Department, Vancouver Island University, Nanaimo, British Columbia Canada. Retrieved from: http://wwwviu.ca/earthscience/fossilsofvancouverisland.pdf. July 14, 2021.

Kemp TS (2005) *The Origin and Evolution of Mammals*. Oxford University Press, Oxford, United Kingdom. p. 217.

Kennedy MP, Moore GW (1971) Stratigraphic relations of Upper Cretaceous and Eocene formations, San Diego coastal area, California. *American Association of Petroleum Geologists Bulletin* 55(5): 709–722.

Koch PL, Barnosky AD (2006) Late Quaternary extinctions: State of the debate. *Annual Review of Ecology, Evolution, and Systematics* 37: 215–250.

Konrad K, Koppers AAP, Steinberger B, Finlayson VA, Konter JG, Jackson MG (2018) On the relative motions of long-lived Pacific mantle plumes. *Nature Communications* 9(1). DOI: 10.1038/s41467-018-03277-x.

Krause J, Dear PH, Pollack JL, Slatkin M, Spriggs H, Barnes I, Lister AM, Ebersberger I, Pääbo S, Hofreiter M (2006) Multiplex amplification of the mammoth mitochondrial genome and the evolution of Elephantidae. *Nature* 439: 724–747.

Kurtén B (1968) *Pleistocene Mammals of Europe*. Transaction Publishers, Piscataway, New Jersey. pp. 170–177.

Lister A, Bahn P (2007) *Mammoths: Giants of the Ice Age*. Frances Lincoln LTD, London, UK. p. 23.

Lister AM, Sher AV (2015) Evolution and dispersal of mammoths across the Northern Hemisphere. *Science* 350: 805–809.

Livingston VE (1959) *Fossils in Washington*. Division of Mines and Geology, Bulletin No. 33. Department of Conservation and Development, State of Washington, Tacoma, Washington.

Lorenzen ED, Nogués-Bravo D, Orlando L, Weinstock J, Binladen J, Marske KA, Ugan A, Borregaard MK, Gilbert MTP, Nielsen R, Ho SYW, Goebel T, Graf KE, Byers D, Stenderup JT, Rasmussen M, Campos PF, Leonard JA, Koepfli K-P, Froese D, Zazula G, Stafford Jr TW, Aaris-Sørensen K, Batra P, Haywood AM, Singarayer JS, Valdes PJ, Boeskorov G, Burns JA, Davydov SP, Haile J, Jenkins DL, Kosintsev P, Kuznetsova T, Lai X, Martin LD, McConald HG, Mol D, Meldgaard M, Munch K, Stephan E, Sablin M, Sommer RS, Sipko T, Scott E, Suchard MA, Tikhonov A, Willerslev R, Wayne RI, Cooper A, Hofreiter M, Sher A, Shapiro B, Rahbek C, Willerslev E (2011) Species-specific responses of Late Quaternary megafauna to climate and humans. *Nature* 479: 359–364.

Ludvigsen R, Beard G (1997) *West Coast Fossils: A Guide to the Ancient Life of Vancouver Island*. Harbour Publishing, Madeira Park, British Columbia, Canada.

MacDonald JH, Schoonmaker A (2017). Evidence of a late Jurassic ridge subduction event: Geochemistry and age of the Quartz Mountain stock, Manastash inlier, central Cascades, Washington. *Journal of Geology* 125(4): 423–438.

Malhi Y, Doughty CE, Galetti M, Smith FA, Svenning J-C, Terborgh JW (2016) Megafauna and ecosystem function from the Pleistocene to the Anthropocene. *Proceedings of the National Academy of Sciences of the United States of America* 113: 838–846.

Marshall M (2010) . Retrieved from: https://www.newscientist.com/article/dn18949-the-history-of-ice-on-earth/. September 10, 2020.The History of Ice on Earth The New Scientist London, United Kingdom.

Martin PS (1984) Prehistoric overkill: The global model. In: Martin PS, Klein RG (editors). *Quaternary Extinctions: A Prehistoric Revolution*, pp. 364–403. University of Arizona Press, Tucson, Arizona.

Martin JE (2008) Hemphillian rodents from northern Oregon and their biostratigraphic implications. *Paludicola* 6(4): 155–190.

Mayr G, Goedert JL (2017) Oligocene and Miocene albatross fossils from Washington State (USA) and the evolutionary history of North Pacific Diomedeidae. *The Auk* 134(3): 659–671.

McLellan RD (1927) *The Geology of the San Juan Islands*. Doctoral Thesis. In: University of Washington Publications in Geology, 1916-1924, Volume 2, University of Washington Press, Seattle, Washington.

Millar CI, Woolfenden WB (2016) Ecosystems Past: Vegetation prehistory. In: Mooney H, Zavaleta E (editors). *Ecosystems of California*, pp. 131–154. University of California Press, Berkeley, California.

Moorbath S, Sigurdsson H, Goodwin R (1968) K-Ar ages of the oldest exposed rocks in Iceland. *Earth and Planetary Science Letters* 4(3): 197–205.

Moore EJ (1963) *Miocene Marine Mollusks from the Astoria Formation in Oregon.* Geological Survey Professional Paper 419, Geological Survey, US Department of the Interior, Washington DC.

Mullineaux DR (1970) Geology of the Renton, Auburn, and Black Diamond Quadrangles, King County, Washington. *Geological Society Professional Paper* No. 674, 24–26.

Murray M (1974) *Hunting for Fossils: A Guide to Finding and Collecting Fossils in All 50 States.* Collier Books, Springfield, Ohio.

Mustoe GE, Carlstad CA (1995) A late Pleistocene brown bear (*Ursus arctos*) from Northwest Washington. *Northwest Science* 69(2): 106–113.

Mustoe GE, Gannaway WL (1997) Paleogeography and paleontology of the early Tertiary Chuckanut Formation, Northwest Washington. *Washington Geology* 25(3): 3–18.

Mustoe GE, Leopold EB (2014) Paleobotanical evidence of the Post-Miocene uplift of the Cascade Range. *Canadian Journal of Earth Sciences* 51(8). DOI: 10.1139/cjes-2 013-0223.

Nesbitt L, Scotchmoor J (2010) *Washington, US.* The Paleontology Portal, UC Museum of Paleontology, University of California Berkeley, Berkeley, California. Retrieved from: www.paleoportal.org. July 14, 2021.

Newham E, Benson R, Upchurch P, Goswami A (2014) Mesozoic mammaliaform diversity: The effect of sampling corrections on reconstructions of evolutionary dynamics. *Paleogeography, Paleoclimatology, Paleoecology* 412: 32–44.

Nordt L, Atchley S, Dworkin S (2003) Terrestrial evidence for two greenhouse events in the latest cretaceous. *GSA Today* 13(12): 4–9. DOI: 10.1130/1052-5173. February 16, 2021.

Nyborg TG, Vega FJ, Filkorn HF (2003) New late cretaceous and early Cenozoic decapod crustaceans from California, USA: Implications for the origination of taxa in the eastern North Pacific. *Contributions to Zoology* 72: 165–168.

Oberbauer T (2018) *Prehistoric San Diego County, Part 2.* California Native Plant Society and San Diego Natural History Museum, San Diego, California. February 27, 2021.

O'Dea A, Lessios HA, Coates AG, Eytan RI, Restrepo-Moreno SA, Cione AL, Collins LS, de Queiroz A, Farris DW, Norris RD, Stallard RF, Woodburne MO, Aguilera

O, Aubry MP, Berggren WA, Budd AF, Cozzuol MA, Coppard SE, Duque-Caro H, Finnegan S, Gasparini GM, Grossman EL, Johnson KG, Keigwin LD, Knowlton N, Leigh EG, Leonard-Pingel JS, Marko PB, Pyenson ND, Rachello-Dolmen PG, Soibelzon E, Soibelzon L, Todd JA, Vermeij GJ, Jackson JB (2016) Formation of the Isthmus of Panama. *Science Advances* 2(8): e1600883.

Peecock BR, Sidor CA 2015) The first dinosaur from Washington State and a review of Pacific coast dinosaurs from North America. *PLoS One* 10(5): e0127792.

Perri AR, Mitchell KJ, Mouton A, Alvarez-Carretero S, Hulme-Beaman A, Haile J, Jamieson A, Meachen J, Lin AT, Schubert B, Ameen C, Antipina EE, Bover P, Brace S, Carmagnini A, Carøe C, Samaniego Castruita JA, Chatters JC, Dobney K, dos Reis M, Evin A, Gaubert P, Gopalakrishnan S, Gower G, Heiniger H, Helgen KM, Kapp J, Kosintsev PA, Linderholm A, Ozga AT, Presslee S, Salis AT, Saremi NF, Shew C, Skerry K, Taranenko DE, Thompson M, Sablin MV, Kuzmin YV, Collins MJ, Sinding M-KS, Gilbert MTP, Stone AC, Shapiro B, Van Valkenburgh B, Wayne RK, Larson G, Cooper A, Frantz LAF (2021) Dire wolves were the last of an ancient New World canid lineage. *Nature* 591: 87–91.

Peterson GL, Abbott PL (1979) Mid-Eocene climatic change, Southwestern California and Northwestern Baja California. *Paleogeography, Paleoclimatology, Paleoecology* 26: 73–87.

Phillips WM, Walsh TJ, Hagen RA (1989) Eocene transition from oceanic to arc volcanism, southwest Washington, In: Muffler LJ, Weaver CS, Blackwell DD (editors). *Proceedings of Workshop XLIV: Geological, Geophysical, and Tectonic Setting of the Cascade Range*, pp. 199–256. US Geological Survey, Washington DC.

Prothero DR, Armentrout JM (1985) Magnetostratigraphic correlation of the Lincoln Creek Formation, Washington: Implications for the age of the Eocene/Oligocene boundary. *Geology* 13(3): 208–211.

Roach J (2003) *Hot Spot That Spawned Hawaii Was on the Move, Study Finds*. National Geographic News, Washington DC.

Sandom C, Faurby S, Sandel B, Svenning J-C (2014) Global late Quaternary megafauna extinctions linked to humans, not climate change. *Proceedings of the Royal Society B: Biological Sciences* 281: 1–9.

Sayce K (2012) *Some Interesting Local Fossils Come in Tidy, Round Rock Packages*. Chinook Observer, Long Beach, Washington. Retrieved from: www.chinookobserver.com/opinion/some-interesting-local-fossils-come-in-tidy-round-rock-packages/article_fefb8015-dba9-5e3e-94eb-3c36e2cb633d.html. July 17, 2021.

Sharpe WD, Clague DA (2006) 50-Ma initiation of Hawaiian-emperor bend records major change in Pacific plate motion. *Science* 313(5791): 1281–1284.

Smith M, Manchester SR (2018) Nut of *Juglans bergomensis* (Balsamo Crivelli) Massalongo in the Miocene of North America. *Acta Paleobotanica* 58(2): 199–208.

Spaulding M, O'Leary MA, Gatesy J (2009) Relationships of Cetacea (Artiodactyla) among mammals: Increased taxon sampling alters interpretations of key fossils and character evolution. *PLoS One* 4(9): e7062.

Stearley RF, Smith GR (2016) I. Salmonid fishes from Mio-Pliocene lake sediment in the western Snake River plain and the Great Basin. In: Stearley RF, Smith GR, Fishes of the Mio-Pliocene western Snake River plain and vicinity. Miscellaneous Publication, Publications of the Museum of Zoology, University of Michigan No.204, University of Michigan, Ann Arbor, Michigan, October 14, 2016.

Stuart AJ, Kosintsev PA, Higham TFG, Lister AM (2004) Pleistocene to Holocene extinction dynamics in giant deer and woolly mammoth. *Nature* 431: 684–689.

Surovell TA, Pelton SR, Anderson-Sprecher R, Myers AD (2016) Test of Martin's overkill hypothesis using radiocarbon dates on extinct megafauna. *Proceedings of the National Academy of Sciences USA* 113: 886–891.

Tabor RW, Cady WM (1978) *Geologic Map of the Olympic Peninsula, Washington.* US Geological Survey, Washington DC. Retrieved from: https://pubs.er.usgs.gov/publication/i994. July 14, 2021.

USGS (2004) *North Cascades Geology.* USGS, Department of the Interior, Washington DC. Retrieved from: http://geomaps.wr.usgs.gov/parks/noca/nocageol1.html. July 19, 2021.

USGS (ND) *Miocene-Pliocene Marine Rocks.* US Geological Survey, US Department of the Interior, Washington DC. Retrieved from: https://mrdata.usgs.gov/geology/state/sgmc-unit.php?unit=WAMIPO%3B0. July 18, 2021.

van der Valk T, Pečnerová P, Díez-del-Molino D, Bergström A, Oppenheimer J, Hartmann S, Xenikoudakis G, Thomas JA, Dehasque M, Sağlıcan E, Fidan FR, Barnes I, Liu S, Somel M, Heintzman PD, Nikolskiy P, Shapiro B, Skoglund P, Hofreiter M, Lister AM, Götherström, Dalén L (2021) Million-year-old DNA sheds light on the genomic history of mammoths. *Nature* 591: 265–269. DOI: 10.1 038/s41586-021-03224-9.

Walsh SL (1996) Middle Eocene mammalian faunas of San Diego County, California, In: Prothero DR, Emry RJ (editors). *The Terrestrial Eocene-Oligocene Transition in North America.* Cambridge University Press, Cambridge, United Kingdom.

Weaver CE (1912) *A Preliminary Report on the Tertiary Paleontology of Western Washington.* City of Olympia, Washington.

Weaver CE (1937) *Tertiary Stratigraphy of Western Washington and Northwestern Oregon.* Washington University-Seattle Publications in Geology, Washington. p. 266.

Weishampel DB, Dodson P, Osmólska H (2004) Dinosaur distribution (Late Cretaceous, North America). In: Weishampel DB, Dodson P, Osmólska H (editors). *The Dinosauria*, pp. 574–588. University of California Press, Berkeley.

Westby (ND) *Columbia River Basalt Group stretches from Oregon to Idaho.* David A. Johnston Cascades Volcano Observatory, US Geological Survey, Vancouver, Washington. Retrieved from: https://www.usgs.gov/observatories/cascades-volcano-observatory/columbia-river-basalt-group-stretches-from-oregon-to-idaho. July 17, 2021.

Wilson JT (1963) A possible origin of the Hawaiian Islands. *Canadian Journal of Physics* 41(6): 863–870.

Woodring WP (1938) *Lower Pliocene Mollusks and Echinoids for the Los Angeles Basin, California and Their Inferred Environment.* Geological Survey, Department of the Interior, Washington DC.

Wroe S, Field J (2006) A review of the evidence for a human role in the extinction of Australian megafauna and an alternative interpretation. *Quaternary Science Reviews* 25: 2692–2703.

WSDNR (2021a) *Puget Lowland.* Washington State Department of Natural Resources, Tacoma, Washington. Retrieved from: www.dnr.was.gov/program-and-services/geology/explore-popular-geology/geologic-provinces-washington/puget-lowland#glacial-features/. July 13, 2021.

WSDNR (2021b) *Olympic Mountains.* Washington State Department of Natural Resources, Tacoma, Washington. Retrieved from: www.dnr.was.gov/program-and-services/geology/explore-popular-geology/geologic-provinces-washington/olympic-mountains/. July 19, 2021.

WSL (2019) Designating the Suciasaurus rex as the official dinosaur of the state of Washington, HB 2155 – 2019-20. Washington State Legislature, Tacoma, Washington.

6 Humans Arrive

For the last few thousand and especially the last few hundred years, the major force for change in the Puget Sound region has been humans. The massive tectonic, volcanic, glacial, and hydrological forces worked for millions of years to build the primary structure of the region. Humans cannot match that impact. However, humans have certainly had a significant influence on the shorelines and rivers, air, and animals, and they have done it in an amazingly short time.

NATIVE AMERICANS

EARLIEST HUMANS AND THEIR LIVES

The first humans in the Puget Sound region arrived long ago. The earliest migrations of modern humans into the Americas are complicated and not completely settled (Waters, 2019). The dates are still a matter of some controversy, but they arrived at least 12,000 years ago and possibly much earlier. The traditional view holds that humans migrated from Siberia to Alaska and spread out around the Americas. They crossed a land bridge called Beringia from Asia about 30,000 years ago (NPCC, nd). These were the "Clovis People," named for a town in New Mexico where their remains were first found. Yet, there is little evidence for them in the Northwest. The land bridge likely appeared and disappeared as sea levels rose and fell with the different Ice Ages, and so, humans may have had multiple opportunities to migrate. During the last Ice Age, the sea levels decreased significantly and opened an overland route into North America.

The Clovis model fell apart as more evidence was found to support a model in which early humans followed the shoreline into America. The newer theory puts more weight on a coastal strategy for the migration. It gained popularity with a particular Y chromosome mutation that occurred in Siberia about 15,000–18,000 years ago. Native Americans share this mutation, and that observation suggests that the colonization of the Americans had at least one wave about that time.

Many sites have been found along the shores, and many more were lost over time to sea level rise, erosion, and tectonic movements (Braje et al., 2017). Nevertheless, sites as far south as Monte Verde in Chile show human habitation at least 14,000 years ago and maybe as much as 16,000–18,000. A stone blade found offshore of Haida Gwaii in British Columbia was at least 10,000 years old. Additional artifacts have been located near the mouth of the Columbia River that are perhaps 1000 years younger. Thus, Native Americans were certainly in the area of the Puget Sound at about that time. By about 3000 years ago, they were preserving salmon by smoking.

DOI: 10.1201/9780429487439-6

Native American tribes in the Puget Sound region are commonly called the Coast Salish people and include the Suquamish, Duwamish, Nisqually, Snoqualmie, and Muckleshoot (Ilalkoamish, Stuckamish, and Skopamish) (Watson, 1999). The population number is unknown, but estimates are that tens of thousands of Native Americans lived in the region (Watson, 1999). Each extended family lived in a large rectangular house made of split cedar boards on a post and beam frame. They used the resources available to them. Cedar bark was used for clothing and baskets.

Their diet included a number of animal and plant foods. The seashore provided many resources, such as salmon, smelt, eulachon, and herring (Quinn, 2010). Shellfish were taken from the beaches and mud flats. They caught salmon during the annual runs and preserved the flesh by smoking or drying. In addition, they hunted elk, deer, seals, bear, ducks, and other prey. The plants included huckleberries, blackberries (*Rubus ursinus*), and the roots of the sunflower (*Balsanwrhiza deltoidei*) and chocolate lily, (*Fritillaria lanceolata*), and kamas or Indian hyacinth) (*Camassia esculenta*). They made a flour with bracken ferns (*Pteridiuin aquilinum*) (Dunwiddie et al., 2014). Bunchgrass (*Festuca idahoensis*) was important for forage for deer. They also ate acorns, especially those from the Oregon white oak (*Quercus garryana*) (Anderson, 2007). Acorns can be stored for a year or more. They are prepared by removing the shells and pounding into a flour. Then it must be leached to remove the bitter tannins. Then it can be cooked as mush, soup, paddies, cakes, bread, and more. They were generally roasted, steamed, or boiled. Hot rocks were used for much of the cooking (Thoms, 2009). A large oven-like site in the Chester Morse Reservoir near Seattle was dated to about 8500 years ago. Examination of other sites showed food remains, such as seeds, kamas, bracken ferns and deer and elk bones.

The prairies in the southern part of the Puget Sound appeared in a particularly warm and dry period 4500–9500 years ago. Those areas featured many plants that were important to the survival of the people. Native Americans used fire to clear trees and shrubs (Boyd, 1999) on those prairies to encourage the growth of those plants. Each year the Salishan people burned the prairie before the autumn rains would begin. Also, by burning in August, they could ensure that a second crop of grasses would be available to feed the deer that were important to them.

Early humans used fire in what has been called "fire-stick farming," in which fire is used to clear ground, kill vermin, and regenerate plant food. It could open up woodlands so that more desirable plants could move in (Pausas and Keeley, 2009). Most archaeologists believe that Native Americans had stable foraging economies, and so, they did not need to develop agriculture (Bettinger, 2015).

Further up the Cascade Mountains, the Native Americans also used fire to clear areas. The burning encouraged the growth of the black mountain huckleberry (*Vaccinium membranaceum*). The patches were burned after the berry harvest each year. The fire also prevented the growth of shrubs and trees. In addition to the huckleberries, the burning support beargrass (*Xerophyllum tenax*), and bearberry or kinnikinnick (*Arctostaphylos uva-urst*).

Unstable forests are more susceptible to fire. Crausbay et al. (2017) examined pollen left by ancient forests in the Pacific Northwest over the last 14,000 years. They found that forests during climate change were not resilient to fire, but those in periods of stasis were.

EXTINCTION OF LARGE MAMMALS

About 50,000 years ago, near the end of the Late Quaternary, many large mammals in North America and Eurasia began to be lost (Barnosky et al., 2004). The reasons are unclear but two causes have emerged as the most likely. North America had more extensive climate change and lost more mammals than Eurasia. The disappearance of the mammals coincided with the arrival of humans in those regions. Those animals included mammoths, mastodons, giant beavers, and others.

There is evidence for two models of extinction. Martin (1967) in their his "overkill hypothesis, suggested that Native Americans hunted those mammals to extinction. Sandom et al. (2014) completed an analysis of all large mammals (greater than 10 kg) that became extinct between the last Interglacial period (132,000 years ago) and the late Holocene (1000 years ago). They found that the loss of the animals was strongly tied to the arrival of humans. A correlation of the timing of 42 archaeological sites and genetic and climatic evidence also shows that humans arrived before the Last Glacial Maximum (26.5 to 19,000 years ago) but probably dispersed during a warming period (Greenland Interstadial) (Becerra-Valdivia and Higham, 2020). Smith et al. (2018) found similar evidence of human involvement. Still not everyone agrees with the overkill hypothesis (Grayson and Meltzer, 2003). The evidence seems to indicate that only the mammoth and the mastodon were hunted. The rest were not. Also, temperatures fluctuated greatly near the end of the Pleistocene. The controversy is unlikely to be resolved soon, but Stuart (2015) strikes a middle ground by implicating both human and climate factors in the loss of the great mammals. Lorenzen et al. (2011) weighed the evidence by looking at the genetic diversity of the six large herbivores: woolly rhinoceros (*Coelodonta antiquitatis*), woolly mammoth (*Mammuthus primigenius*), horse (wild *Equus ferus* and living domestic *Equus caballus*), reindeer/caribou (*Rangifer tarandus*), bison (*Bison priscus/Bison bison*), and musk ox (*Ovibos moschatus*). They concluded that the climate was a major driver of the changes. It can probably explain the loss of the Eurasian musk ox and woolly rhinoceros. However, both climate and seem to be involved in the extinction of others, including Eurasian steppe bison and wild horse.

EARLY EUROPEANS

Beginning in the 18th century, Europeans began visiting the Puget Sound region. The first European to come to Washington state was Juan Perez in 1774. A year later Don Bruna de Heceta landed near the mouth of the Quinault River and claimed the land for Spain and gave it its name. In 1787, English captain Charles William Barkley first discovered the Strait of Juan de Fuca. In 1792, George Vancouver claimed the Sound for Britain and named it in honor of Peter Puget, his lieutenant.

The attraction of the area initially was the fur trade. The most desirable furs were those of the sea otter (*Enhydra lutris kenyoni*), which have thicker fur than many other mammals. Several European nations wanted the furs. The Russians had a great interest in the fur trade on the West Coast of the current United States. Their most southern colony was at Fort Ross in Northern California. The Spanish claimed California as far north as San Francisco as a bulwark against Russian expansion. At the same time, the British were also staking a claim to the region. More settlers coming by the Oregon Trail moved north into the area. Before the hunting began, the population was estimated at 150,000–300,000. However, after years of hunting, those animals were nearly extinct. With the hunting of beaver, the number of beaver dams were greatly reduced. Young coho salmon depend on beaver ponds early in life, and their numbers were also reduced by the hunting of beaver (Beechie et al., 2001; Pollock et al., 2004).

The first European settlement was at New Market (near current-day Olympia) in the southern Puget Sound. Others established Fort Nisqually near the current city of Dupone, WA, just south of Tacoma. It was just a farm and fur-trading post and a subsidiary of the Hudson's Bay Company. Fort Nisqually exported livestock and crops to Russian Alaska, Hawaii, and Northern California. In 1845, there were over 2000 cattle and 6000 sheep. The Fraser Gold Rush in British Columbia brought more people through the area. But the main industry was lumber. Lumber was harvested in the Puget Sound area and shipped to supply the growing needs of San Francisco. Lumbering began with axes and horse teams working along the water. That cleared a lot of land that would later be used for other industries.

European settlers first arrived in what would become Seattle in 1851 (Seattle, 1995). They established the city in the Pioneer Square district. The main industry was a lumber mill owned by Henry Yesler that opened in 1853. Much of that lumber was shipped to San Francisco to support the growth there caused by the Gold Rush. By 1869, the town had more than 2000 people. The railroad arrived in Seattle and Tacoma in the 1880s, and the towns boomed after that. By the end of that decade, Seattle was growing at a rate of 1000 new people per month. In 1897, gold was discovered on the Klondike River in the Yukon Territory, and Seattle prospered as one of the paths to the goldfields. By 1909, the population was nearly 240,000.

As the population grew into the millions in the 20th century, the major industries grew from lumber, coal, and shipping to include military bases, shipyards, and aircraft manufacturing (Gregory, 2017). In 1950, the population of the Seattle-Tacoma-Bellevue metropolitan area was just under 800,000. By 2020, it was nearly four million. The rapid influx of people into the area was accompanied by a massive amount of building of housing, commercial interests, and support infrastructure (e.g., roads, bridges). All of this greatly affected the environment of the Puget Sound region. For example, the rivers were changed by dams, diversions, and land use. These blocked routes used for migration and changed the characteristics of the water and sediment (e.g., chemistry, physics, amounts, timing) that was delivered to the estuaries.

Over the last 150 years, most of the extractable natural resources have been depleted (Quinn, 2010). About 60% of the old-growth forest has been cut. Nearly a quarter of the forest lands have been used for development. The salmon runs are greatly reduced, and 80% of the wetlands have been drained. The shoreland was developed for port activities and filled for development. Overall, water quality throughout the region has declined. A number of species have been impacted, including the Puget Sound Chinook salmon (*Oncorhynchus tshawytscha*), steelhead salmon (*Oncorhynchus mykiss*), and killer whales (*Orcinus orca*).

The Puget Sound region is distinct from much of the rest of the United States. It was a settled lake and its weather systems are different. To study the use of fossil fuels and biomass, Kuo et al. (2011) looked at the deposition of black carbon in the Sound and in Hood Canal. The Sound region has grown rapidly, but the Hood Canal is still sparsely populated. They attempted to reconstruct 250 years of deposits of two types of black carbon: charcoal and soot or graphitic black carbon. The peak amounts were in the 1940s for the Puget Sound and the 1970s for Hood Canal. They also concluded that more softwoods were burned in Washington than on the East Coast and that changes in climate also tracked well with the number of wildfires.

BUILDING AND DEVELOPMENT

The Puget Sound region has developed into one of the largest metropolitan areas. Over three million people live there, and a great deal has been done to alter the natural landscape. The occupies 2600 km^2, according to the US Census Bureau. Kings County, which contains Seattle, has 2400 km of roads and 182 bridges (Kings County, 2021). The residents also need a healthy water supply and an efficient sewage disposal system, and so the county has five sewage treatment plants (Kings County, 2020). The West Point and South plants handle 90 million gallons/day in the dry season and up to 300 million/day in the storm season), Brightwater handles 36 and 54 million in the dry and storm seasons, respectively. Vashon and Carnation together handle nearly 700,000 gallons.

INDUSTRY

A number of industries also sprang up in the region. They provided income for the people; however, in that era, many were not as environmentally conscious as they are today. Their legacy remains a problem for those living there. The first industry was logging (Soundkeeper, nd). It began in 1853, and by 1883, nearly all the timber had been removed. Huge amounts of sediment were washed into streams, and a great deal of habit was destroyed. In the 1920s, pulp mills started discharging pollutants, including dioxins, into the Sound. The shellfish industry was devastated. In 1888, the Ryan Smelter began refining lead (Gilbert, 2019). Early in the 20th century, it was converted to produce copper. It featured an enormous smokestack that was considered a wonder. However, at 571 feet, it only allowed the pollution of arsenic and other heavy metals to spread over an even larger area. The SO2 emissions resulted in a horrible smell, and finally, in 1950, the company built an acid plant that used the SO2 to produce sulfuric acid. Yet, it did not solve the problems of pollution, which had spread over 2500 km^2.

Filling and Dredging

Tacoma tide flats are part of the Puyallup-White River Watershed that extends from Mount Rainer to Commencement Bay (Tacoma Tideflats, nd). Beginning in the 1930s, the tide flats began to be filled. The combination of major rail lines and a deep-water port enabled the lumber industry to grow. The ready supply of wood also spurred growth of the paper and wood products industry and the chemical industry to support them. The tide flats provided a large amount of seemingly wasted land for expansion. To control its flow and prevent flooding the Puyallup River was diked and more areas of the estuary were filled. With the industrial activity and poor control of waste, pollution increased and the water quality degraded. Dangerous chemicals were released into the soil and ground-water. The air quality was also poor.

Commencement Bay has been greatly altered by human activities (Patterson, nd). The Bay estuary originally featured a large freshwater floodplain that served as a habitat for juvenile salmon. In the last quarter of the 19th century, lumber mills were built on the tide flats. More changes to the Bay began in the early 1900s. The many streams of the Puyallup River were channeled to create the eight waterways there now. Warehouses and grain storage facilities grew in the early 20th century. All of this was accomplished with considerable dredging and filling of the estuary. Many native species were affected, including the salmon.

In like manner, the Duwamish River begins as the Green River in the Cascade Mountains, joins the Black and White Rivers, and then meanders down through mudflats and marshes (Richards et al., 2012). A flood changed the course of the White River so that it now flows into the Puyallup River. The Black River was lost after one of its main tributaries was diverted into Lake Washington. In the early 20th century, the growth of industry caused much of the lower section of the river to be dredged and straightened. Overtime, the estuary was changed into the Lower Duwamish Waterway. Dredged material was used to build Harbor Island and fill parts of the old channel. The Waterway continues to be dredged today and is a major shipping route with all of the infrastructure required for that purpose (e.g., bulkheads, piers, wharves, structures).

The original topography of Seattle has changed dramatically. Hills were leveled and used to fill low-lying sections (e.g., the current central waterfront). An estimated 45 million tons of material was moved in dozens of projects. One of the largest in 1908–1911, millions of gallons of water per day were used to erode away Denny Hill. Another regrade at this site in 1929 completed the job. In another project, the 350-acre Harbor Island was constructed in 1909. About 18 million cubic meters of material was taken from the Jackson and Dearborn street regrades and dredged material that was dumped at the mouth of the Duwamish River.

Sewage and Solid Waste Disposal

By 1890, Seattle's population relied on unreliable water and sewage system. Cholera and typhoid were serious concerns. In 1889, much of the city was destroyed by a fire that was exacerbated by a lack of water to fight it. To remedy these situations, the city built a pipeline to transport water from the Cedar River at Landsburg Dam to the city. Over time, the Cedar River Municipal Watershed took over and protected the watershed from further development. As a result, in 1901, water from the Cedar River in the Cascade Mountains was diverted for the first time into a pipe at Landsburg Dam for the 46-km trip to the city.

The city of Seattle began contracts for solid waste collection in 1911. Early on, solid waste was handled by dumping it around the city in tide flats and other areas. Some property owners welcomed the material to build up their land for development. However, by the mid-1920s (Seattle-King, 1984), 16 dumps were scattered around town, and the public was becoming much less tolerant of using the city as a dumping ground. In 1938, the engineering department took responsibility for those contracts from the health department, and in 1943, they began using a single contractor. In 1966, two landfills and two transfer stations were established.

The Duwamish River has been greatly affected by human activities (Ott, 2016). The river has been subjected to many measures to straighten and control it. It has also been used as an easy and cheap disposal site for human and industrial wastes, especially in the 1940s and 1950s.

The challenges of dealing with sewage increase with increasing population density. Seattle originally used open ditches and cesspools to take care of sewage, but by the late 19th century, the population had exceeded the ability of those measures. In 1883, the city built its first wooden-box sewers, and in 1884, they required residents to connect to that system.

In the early 20th century, the city began to construct a sewage system that carried both waste and stormwater in the same system. That system carried the material without processing to various bodies of water, including Lake Washington, Lake Union, the Duwamish River, Elliott Bay, and the Puget Sound. Other towns followed Seattle's lead. And for a while, this system worked. The carrying capacity of the rivers was sufficient, and most of the material was biodegradable. However, population growth soon outstripped the capacity of the rivers, and a new solution was needed. A more modern system was built, but raw sewage still entered the waterways, especially during storms, and from other cities with less sophisticated systems along the rivers. By 1945, all beaches in Seattle were unsafe for bathing due to high levels of coliform bacteria from raw sewage.

Sewage was not the only pollutant entering the Duwamish River. Industrial concerns released chemicals (e.g., chromic acid, cutting oils, sodium nitrate), paint, glue, caustic soda, charcoal, and more onto the land and into the water.

COMPLEX SYSTEMS

The interactions of humans with their natural environment are complex and not always predictable (Liu et al., 2007). Those interactions and their consequences differ with the extent of development. Changes to the land cover influence biophysical processes, water flow, and biological diversity. Many of the inter-actions are nonlinear and the infrastructure that was developed even 100 years ago can influence the landscape today. For example, the diversity of birds in the Puget Sound region increases nonlinearly when 50–60% of the land is forested. Groundwater takes a long time to accumulate and to replenish, and movements between different parts of an aquifer is extremely slow. Thus, disturbances to the quality of the aquifer water can take a long time to be seen. In the Puget Sound region, in 1991–1999, developed land area increased by 620 km^2 (up 31.5%), and forested areas declined by 714 km^2 (down 10.3%). Fortunately, en-vironmentalists have been active in the region, and the Puget Sound contains one of the last old-growth forests in the United States.

REFERENCES

Anderson MK (2007) *Indigenous Uses, Management, and Restoration of Oaks of the Far Western United States*. National Plant Data Center, US Department of Agriculture, United States.

Barnosky AD, Koch PL, Feranec RS, Wing SL, Shabel AB (2004) Assessing the causes of late Pleistocene extinctions on the continents. *Science* 306: 70–75.

Becerra-Valdivia L, Higham T (2020) The timing and effect of the earliest human arrivals in North America. *Nature* 584: 93–97.

Beechie TJ, Collins BD, Pess GR (2001) Holocene and recent geomorphic processes, land use and salmonid habitat in two north Puget Sound river basins. In: Dorava JB, Montgomery DR, Fitzpatrick F, Palcsak B (editors). *Geomorphic Processes and Riverine Habitat, Water Science and Application,* Vol 4. American Geophysical Union, Washington, D.C.

Bettinger RL (2015) *Orderly Anarchy: Sociopolitical Evolution in Aboriginal California*. University of California Press, Berkeley, CA. pp. 137–138.

Boyd, R. 1999. *Indians, Fire and the Land in the Pacific Northwest*. Oregon State University Press, Corvallis.

Braje TJ, Dillehay D, Erlandson JM, Klein RG, Rick TC (2017) Finding the first Americans. *Science* 358: 592–594.

Crausbay SD, Higuera PE, Sprugel DG, Brubaker LB (2017) Fire catalyzed rapid eco-logical change in lowland coniferous forests of the Pacific Northwest over the past 14,000 years. *Ecology* 98: 2356–2369.

Dunwiddie PW, Alverson ER, Martin RA, Gilbert R (2014) Annual species in native prairies of South Puget Sound, Washington. *Northwest Science* 88: 94–105.

Gilbert SG (2019) *Tacoma Smelter: A Toxic Legacy of Lead and Arsenic Contamination*. Collaborative on Health and the Environment. Retrieved from: https://www.health

andenvironment.org/environmental-health/social-context/history/tacoma-smelter-a-toxic-legacy-of-lead-and-arsenic-contamination.

Grayson DK, Meltzer DJ (2003) A requiem for North American overkill. *Journal of Archaeological Science* 30: 585–593.

Gregory J (2017) *Washington State Migration History 1860–2017.* America's Great Migration Project, University of Washington. Retrieved from http://depts.washington.edu/moving1/Washington.shtml. March 17, 2021.

Kings County (2020) *Waste Water Treatment Division.* Retrieved from: https://www.kingcounty.gov/depts/dnrp/wtd/system/brightwater.aspx. March 18, 2021.

Kings County (2021) *Bridges and Roads Task Force.* Retrieved from: https://kingcounty.gov/depts/local-services/roads/roads-task-force.aspx. March 18, 2021.

Kuo L-J, Louchouarn P, Herbert BE, Brandenberger JM, Wade TL, Crecelius E (2011) Combustion-derived substances in deep basins of Puget Sound: Historical inputs from fossil fuel and biomass combustion. *Environmental Pollution* 159: 983–990.

Liu J, Dietz T, Carpenter SR, Alberti M, Folke C, Moran E, Pell AN, Deadman P, Kratz T, Lubchenco J, Ostrom E, Ouyang Z, Provencher W, Redman CL, Schneider SH, Taylor WW (2007) Complexity of coupled human and natural systems. *Science* 317: 1513–1516.

Lorenzen ED, Nogués-Bravo D, Orlando L, Weinstock J, Binladen J, Marske KA, Ugan A et al. (2011) Species-specific responses of Late Quaternary megafauna to climate and humans. *Nature* 479: 359–364.

Martin PS (1967) *Prehistoric overkill. Pleistocene Extinctions: The Search for a Cause.* Martin PS , Wright HE (editors), pp. 354–403. Yale University Press, New Haven, CT.

NPCC (nd) *Early humans.* Northwest Power and Conservation Council. Retrieved from: https://www.nwcouncil.org/reports/columbia-river-history/firsthumans. March 15, 2021.

Ott J (2016) *Wastewater treatment and the Duwamish River.* HistoryLink.org. Retrieved from: https://www.historylink.org/File/11250.

Patterson K (nd) *History and Ecology of Commencement Bay.* University of Washington. Retrieved from: http://courses.washington.edu/commbay/salmon/commencementbay.html. March 17, 2021.

Pausas JG, Keeley JE (2009) A burning story: The role of fire in the history of life. *BioScience* 59: 593–601.

Pollock MM, Pess GR, Beechie TJ, Montgomery DR (2004) The importance of beaver ponds to coho salmon production in the Stillaguamish River basin. *North American Journal of Fisheries Management* 24: 749–760.

Quinn, T (2010) An environmental and historical overview of the Puget Sound ecosystem. In: Shipman H, Dethier MN, Gelfenbaum G, Fresh KL, Dinicola RS (editors). *Puget Sound Shorelines and the Impacts of Armoring—Proceedings of a State of the Science Workshop*, pp. 11–18. U.S. Geological Survey, Washington.

Richards WH, Koeck R, Gersonde R, Kuschnig G, Fleck W, Hochbichler E (2012) Landscape-scale forest management in the municipal watersheds of Vienna, Austria, and Seattle, USA: Commonalities despite disparate ecology and history. *Natural Areas Journal* 32: 199–207. DOI: 10.3375/043.032.0209.

Sandom C, Faurby S, Sandel B, Svenning J-C (2014) Global late Quaternary megafauna extinctions linked to humans, not climate change. *Proceedings of the Royal Society B: Biological Sciences* 281: 20133254.

Seattle (1995) *Brief History of Seattle. Seattle Municipal Archieves.* Seattle.gov. Retrieved from: https://www.seattle.gov/cityarchives/seattle-facts/brief-history-of-seattle. March 17, 2021.

Seattle-King (1984) *Abandoned Landfill Study in the City of Seattle.* Seattle-King County Department of Public Health. Retrieved from: https://www.kingcounty.gov/depts/

health/environmental-health/toxins-air-quality/~/media/depts/health/environmental-health/documents/toxins/abandoned-landfills seattle 1984 ashx. March 17, 2021.

Smith FA, Elliott Smith RE, Lyons SK, Payne JL (2018) Body size downgrading of mammals over the late Quaternary). *Faculty Publications in the Biological Sciences* 752.

Soundkeeper (nd) *Pollution in Our Waterways*. Puget Soundkeeper. Retrieved from: https://pugetsoundkeeper.org/pollution/#:~:text=Historical%20Sources%20of%20Pollution,Sound%20environment%20was%20through%20logging.&text=The%20first%20pulp%20mills%20in,in%20the%20paper%20bleaching%20process. March 18, 2021.

Stuart AJ (2015) Late Quaternary megafaunal extinctions on the continents: A short review. *Geological Journal* 50: 338–363.

Tacoma Tideflats (nd) *History of Tacoma Tideflats*. Puyallup Watershed Initiative. Retrieved from: https://www.arcgis.com/apps/Cascade/index.html?appid=767eb51aff884d1594542cdc5c52785e. March 27, 2021.

Thoms AV (2009) Rocks of ages: Propagation of hot-rock cookery in western North America. *Journal of Archaeological Science* 36: 573–591.

Watson KG (1999) *Native Americans of Puget Sound: A Brief History of the First People and Their Cultures*. HistoryLink.org. Retrieved from: https://www.historylink.org/File/1506. March 14, 2021.

Waters MR (2019) Late Pleistocene exploration and settlement of the Americas by modern humans. *Science* 365: eaat5447. https://science.sciencemag.org/content/365/6449/eaat5447

7 Puget Sound Today

The Puget Sound of today resulted from the collision of two huge tectonic plates, the subjection of one plate under the other, and then the grinding down of multiple massive glaciers during the Ice Ages. Since then, the rise and fall of sea levels, volcanos, rain and snow, rivers and more have created a beautiful natural bay with an amazingly diverse array of plants and animals. The volcanoes are quiet right now. The Puget Sound is a large estuary with nearly 3800 km of shoreline. The tides bring enormous amounts of water into and out of the Sound every day. The bottom is as rugged as the mountains that surround it.

The Sound covers a large part of Washington state. It is bounded on the east by the Cascade Mountains and on the west by the Olympic peninsula. To the north, it opens into the Strait of Juan de Fuca and then into the Pacific Ocean. The watershed drains over 42,000 km^2. To the south are lowlands. Many rivers flow into the Sound. The largest are the Cedar/Lake Washington canal, Duwamish/Green, Elwha, Nisqually, Nooksack, Puyallup, Skagit, Skokomish, Snohomish, and Stillaguamish Rivers, but thousands more rivers and streams also enter the Sound and the Pacific.

Humans were and still are attracted to the area because of its natural wonders and riches. Unfortunately, people have not always been the best stewards of the environment, and the greatest changes to the Sound for the last hundreds of years have resulted from human activities. Parts of the Sound have been dredged, and the spoils have been used to fill much of the tidal flats for construction. Rivers have been channeled and dammed. An enormous amount of infrastructure has been required for housing, employment, transportation, and entertainment ofthe millions of who now live in the Sound region. Pollution has fouled water in the Sound and the air above it.

Here we will review the current status of the physical Puget Sound, including the land surrounding it, the water in the Bay and the rest of the region, the mudflats and marshlands between the land and water, and finally the climate and air about the region. For a more comprehensive overview of the state of the Puget Sound ecosystem today, we refer the reader to Ruckelshaus et al. (2007).

PEOPLE AND MORE PEOPLE

Seattle has experienced continuous growth for the last century (Balk, 2019). In 1900, the population was 100,000. In 1950, it was 500,000. In 2018, it was 745,000. In fact, Seattle is the fastest-growing city in the nation. In 2016, Seattle passed Baltimore to become the tenth most populous city in the United States (Balk, 2016). The population density is 7962 people/square mile. This is a 10%

DOI: 10.1201/9780429487439-7

increase over 2010. The metropolitan area is also growing rapidly. Just under four million people live in the Seattle-Tacoma-Bellevue Metropolitan Area as of 2019. It is the 15th most populous metropolitan area in the United States, and it gained about 500,000 since 2010. Other large cities include Olympia, Bremerton, Everette, and Bellingham.

Traditionally, the aerospace industry was the most important industry in the region. The Boeing Company, one of the two largest airplane manufacturers in the world, is headquartered in Seattle. Forest products were also a traditionally key industry. But in more recent years, the economy has diversified to include transportation equipment, food processing, advanced computer technology, biotechnology, and more. The Port of Seattle is a major trade center and a very large container cargo port. It is also served by the Seattle-Tacoma International Airport and multiple interstate highways and freeways. Tourism and cruise ships are important to the area. The US military has one major facility in the region: Joint Base Lewis-McChord is the home of a US Army corps and a US Air Force Wing. Together they have about 45,000 active duty and civilian employees.

All of the people who live and work in the Puget Sound region require housing, transportation, entertainment, water, sewage, schools, and more. All of these activities stress the natural environment, and those stresses increase as more people move into or visit the area.

LAND

The land includes mountains, low lands, and tidal flats that surround the Sound. The basic structure of the region has remained the same for millions of years. It was created by the collision of two tectonic plates. The glaciers are long gone, but the Juan de Fuca plate continues to subduct underneath the North American plate, and the Cascade volcanoes remain active from the energy released from the subduction. More recent changes have been due to water, weather, and humans.

EARTHQUAKES AND OTHER MOVEMENTS CAUSED BY PLATE MOVEMENTS

The Puget Sound region sits on the "Ring of Fire" that surrounds the Pacific Ocean and has earthquakes regularly. Significant earthquakes were felt in 1700, 1833, 1869, 1872, 1880, 1939, 1945 magnitude 5.5), and 1946 (magnitude 6.3). The 1700 earthquake created an 11-m tsunami.

The earthquake threat to the region is of three types (Hyndman et al., 2003). Crustal or shallow earthquakes occur along faults near the surface (i.e., in the North American plate) at depths of 1–30 km. Intraplate or deep earthquakes occur at depths of 30–70 km in the oceanic crust and are not much felt at the surface. Subduction zone or megathrust earthquakes begin between the two plates and can be very large and dangerous. Van Wagoner et al. (2002) surveyed the Puget lowlands to create a three-dimensional model of the faults in the region and to better predict crustal earthquakes. They found six distinct basins that are

3–10 km in depth and separated by Crescent formation rocks that reach to within 1–2 km of the surface. The patterns are complex and not easy to interpret.

The most dangerous type of earthquake would be a sudden movement along the entire subduction zone. These great earthquakes, sometimes reaching a magnitude greater than 9.0, have occurred in the past (Nelson et al., 2006). Large earthquakes of this type are known from 1000, 1350, 1700, 2500, 3400, 3800, 4400, and 4900 years ago. They occur about every 500 years.

However, the greatest risk to Seattle is the Seattle Fault Zone, which runs east to west through the city (Seattle, nd). Earthquakes on this fault can be as large as a magnitude 7.5, and they have occurred three or four times in the last 3000 years. Seattle is particularly prone to ground failure or liquefaction during earthquakes. About 15% of the city is in this category. Also, the city has a lot of older masonry buildings that are at great risk. Finally, the city depends on bridges that might be damaged in an earthquake.

Tacoma is at risk for an earthquake too. Another east-west fault runs near that city for about 56 km from Belfair through Vashon Island to Federal Way. Gomberg et al. (2010) ran computer simulations to model a 7.1 magnitude earthquake on this fault. This surface earthquake would cause considerable damage from shaking and land movements. Such an earthquake would likely have many aftershocks, and some of them would cause additional damage. Ground shaking is only one of the hazards of an earthquake. The temblor would probably cause a dangerous tsunami within the Sound. The low-lying areas in the Puyallup River delta would be inundated, and disturbingly, a wave of 4 m would strike within 5 minutes of the earthquake.

OTHER LAND MOVEMENT

Earthquakes are not the only ways that soil and rock move. Movements of the plates cause rise and fall of the surface due to compression and expansion. Mud and silt flow downhill in a manner similar to water. Landslides result from excessive rainfall, earthquakes, and other factors. Mud and other debris flows, and a heavy rain accelerates the actions. Landslides cost lives and damage property.

Landslides

Landslides may seem random events, but they have many common features. Hungr et al. (2014) reviewed the classifications and characteristics of landslides. The sequence of events in landslides can be divided into three phases: pre-failure deformations, failure, and post-failure displacements. They do not always happen at once, and they can be rapid or slow. The signs of a pending landslide include ground cracks, sinking areas, leaks in pipes, tilting poles or fences, new cracks in structures. Rock strength and slope can be combined to create classes of landslide susceptibility (Wilson and Keefer, 1985; Ponti et al., 2008; Wills et al., 2011). Using these measurements, slopes can be rated. On low slopes, landslides are less frequent, regardless of the strength of the rocks. The risk increases with the slope and the weakness of the rocks. For thousands of years, humans have been a major factor in earth movements. Clearing land for crops, housing, and other developments encourage erosion that fills rivers and streams. Deforestation also encourages erosion, and dams have stopped the natural movement of the material (Voosen, 2020).

New imaging technologies, such as the space-based synthetic aperture radar interferometry (InSAR), have significantly improved measurements of surface deformations that are helpful for examining faults, ground movements up and down, landslides, and more. Movement of the plates along the faults can cause the ground to rise or subside as the surface tried to overcome compression or spreading. Downward movements can result from depletion of aquifers and settling of unconsolidated sediments.

Landslides are common in the Puget Sound region, and Seattle has been maintaining records of landslides for many years. Coe et al. (2004) examined 90 years of historical records to model landslides in the Seattle area. They found that about 9% of the city has a 1% or greater chance of a landslide triggered by rain each year. That might not sound like much, but some areas are at even greater risk. One hillside in West Seattle has a 25% chance of a landslide. Slope and geology are the two main factors. Several studies used aerial photographs to identify areas that are at risk for landslides. By using LIDAR, Schulz (2007) expanded the number of areas at risk. Not surprisingly, he found that 80% of the landslides were due to human activities. One key factor in landslides is the erosion of the slope-toes of hillsides. In more recent years, many of those have been shored up, and that action will likely prevent some landslides. However, many others have not been reinforced and remain at risk.

Rainfall seems to be a key element in the initiation of landslides. Most of the shallow slides result from rain. Rotational and translational slides are less frequent. In the winters of 1995–1996 and 1996–1997, the region experienced a large number of slides (Baum et al., 2005). Understanding the relationship between rain and ground slippage is important to managing the risks of landslides. Godt et al. (2006) correlated rainfall and landslide data to create a model of landslides in the Seattle area. They took into account the saturation of the ground before rainfall as well. They suggested that this model might be used to develop an early warning system.

Interestingly, landslides are not always rapid and dramatic. Slow-moving landslides are a problem throughout much of the Puget Sound region and many other areas on the west coast. They rarely cost lives, but they can be extremely destructive to structures and infrastructure. The slides depend on the soil, climate, and earthquake activity of the area, but the mechanisms that initiate and maintain them are not well understood. They often occur in soils rich in clay and rock that are mechanically weak and have high levels of seasonal precipitation.

Underwater Landslides

Landslides do not just occur on dry land. They happen at sea as well, and those landslides can cause extreme damage through tsunamis. Underwater landslides can be related to slides, slumps, reef failures, and earthquakes. Not all produce a tsunami, but some do. Local tsunamis are particularly dangerous because there is little time for a warning.

Shaffer and Parks (1994) documented the aftermath of a landslide onto an intertidal beach near Seattle in 1991. It did not take long for most of the material to be washed into the water. Within 23 days, 77% of the sediment was transferred into the nearshore waters. Once in the water, the material occluded the sunlight into a part of the kelp beds, resulting in different densities of *Costaria costata*, *Laminaria* sp., and *Nereocystis luetkeana*. Thus, landslides can affect plants and animals even underwater.

Coastal Erosion

The Sound has steep coasts and sand and gravel beaches. Waves formed by the wind are limited by the relatively small surface area of the water throughout much of the Sound, and longshore sediment transport is important in the movement of the beach material (Shipman, 2010).

Most of the sand and gravel on a beach actually comes from the cliffs and bluffs that surround the beach (Ecology, nd). Those bluffs are called "feeder bluffs" because they feed the beach. The Puget Sound has 2200 km of beaches with sand or gravel. The erosion is caused by rain, wind, wave action, and human activity, and the sediment varies, depending on the height of the bluffs, the proportion of beach-size sediment in the bluff, and the rate of bluff retreat.

Much of the geomorphology of the Sound region resulted from the glaciers that scoured the region thousands of years ago (See Chapter 4 for a more detailed description). Those glaciers left steep coastlines, complex shorelines, and wave environments, and beaches of sand and gravel. Beaches are influenced by the material that is deposited and the energy of the wave action that plays on them. The evolving beaches affect the plants and animals that live there.

Once the sediment is delivered to the beach, it is up to the waves to transport it to its final destination. Maintaining the beaches involves a careful balance between the removal of sediment by wave action and the addition of new sediment from erosion. That critical balance is easily disrupted by human activities. With their beautiful views of the Sound, the bluffs are popular places to build. Of course, while some human actions accelerate the erosion, and others slow it down. Armoring and development have many effects. Armoring the bluffs with sea walls and other protection not only slows erosion but also reduces the amount of sediment reaching the beaches (Johannessen and MacLennan, 2007). They change the back beach areas with development. They also separate the beach and water from the terrestrial environments, and they protect the bluffs and coast from wave actions. Sobocinski et al. (2010) assessed the effects of armoring on insects and benthic macroinvertebrates in the supratidal zone of Puget Sound beaches. Natural beaches had more oligochaetes and nematodes, talitrid amphipods, insects, and collembolans. Developed beaches had more crustaceans. About a third of the Sound is already armored (Shipman, 2010). Global warming brings another variable to beach maintenance. As temperatures rise, the ice caps and glaciers melt, and sea levels rise. More of the bluffs will then be exposed to wave and storm action at the foot or slope-tow, and the erosion will increase.

Coastal erosion results from marine, subaerial, and human processes. They can also work together. Wind forms and pushes waves against the coast and cause erosion over long times. Rain triggers landslides at on steep portions. The erosion process itself is complicated. Many factors determine the extent and rate, including the rocks composition of the cliff, beach width, elevation, and amounts of precipitation. Careful measurements are required to understand the process in any area. Johannessen (2001) examined beaches and bluffs in King and southern Snohomish Counties in Central Puget Sound. He concluded that the region would be better served by using soft shore protection rather than the hard armoring that is commonly used. Under his recommendations, some of the armoring should be removed and alternative methods should be employed to protect the beaches and the bluffs.

WASTE DISPOSAL

More than four million people live in the Seattle-Tacoma-Bremerton metropolitan area, and that many people create a lot of waste that must be properly disposed of. As metropolitan areas continue to grow, the disposal of trash becomes a significant challenge. Like many areas in the United States, Seattle encourages residents to sort waste into recyclables, compostables, and trash. They also put some teeth into their encouragement: they were the first city to actually fine residents for disposing of food incorrectly. Seattle Public Utilities estimates that every family throws away nearly 200 kg of food each year. The city hopes to increase its recycling and composting to 60% of all waste. That would result in much less waste going to expensive landfills.

Seattle is the first city to actually fine residents for disposing of food incorrectly. Seattle Public Utilities estimates that every family throws away nearly 200 kg of food each year. The city hopes to increase its recycling and composting to 60% of all waste. That would result in much less waste going to expensive landfills. Thus far, the city is a bit below its target level. Offenhuber et al. (2012) placed GPS sensors on 2000 discarded items to track them to their final locations. They found that electronic and household hazardous waste have a longer track than other types of waste. Thus, their transportation becomes an important consideration.

Seattle prioritizes the prevention of waste so that less material is sent to expensive landfills (Seattle, 2018). The program has been quite successful. The per capita waste generation rate is now slightly less than a kilogram. They are now focusing on not producing the waste in the first place as an additional strategy. In 2018, Seattle diverted 450,000 tons of material. That is nearly 6000 tons less than in 2017, but the total amount sent to landfills actually decreased slightly, and that is the real goal. In fact, they achieved this result even though the population continued to grow and economic activity was strong. The exact reason for the drop was not completely clear.

Also in 2018, the disposal of plastics was made far more complicated by a change in policy by China (De Freytas-Tamura, 2018). For many years, China had accepted recycled plastic and paper. In 2016, they imported and processed half of the world's waste paper, metals, and plastic (about 7.3 million tons). However, their China Blue Sky policy ended that. China now bans 24 types of solid waste (e.g., unsorted paper, polyethylene terephthalate) and sets tighter standards for impurities in the waste. Their new policies will require Seattle and all solid waste organizations to revise their own efforts.

King County has only one active landfill, the Cedar Hills Regional Landfill (King, 2020). The Landfill is owned by the County and located in Maple Valley, about 20 miles southeast of Seattle. It has operated for over 50 years. Trash is brought to the landfill from King County's eight urban transfer stations and two rural drop boxes. The landfill receives 800,000 tons of solid waste a year. Plans are in place for the landfill to continue operations for another 20 years. The County also oversees nine closed landfills (i.e., Cedar Falls, Duvall, Enumclaw, Hobart, Houghton, Puyallup/Kit Corner, and Vashon). These are monitored for ground water, surface water, and wastewater, as well as landfill gas.

The County also partners with Bio Energy LLC to capture landfill gas from decomposing organic material at the Cedar Hills facility. Methane is collected through a network of pipes throughout the landfill, separated from other gases, and transferred to a natural gas pipeline for energy use.

WATER

Puget Sound covers about 2330 km^2. It is connected to the Strait of Juan de Fuca, which is 160 km long. The Sound is an estuary. Saltwater from the Pacific Ocean is one important factor in the characteristics of the Sound water. Freshwater runoff from rivers and streams and wastewater outflows are another.

The arrival of Americans brought huge changes to the Sound, and many of those changes were not healthy for the natural environment. In 1853, Henry Yesler built a steam-powered logging mill (Soundkeeper, nd). Within a few years, all of the trees around the Sound had been logged. Without the trees to hold it back, large amounts of sediments washed into the Sound and affected the fish and other organisms. In the 1920s, pulp mills started up. The paper bleaching process resulted in the release of toxins, such as dioxins, that damaged the shellfish industry in the southern part of the Sound. A paper plant in Snohomish County operated for decades before it closed in 2012 (Sheets, 2012). Dioxins are leaching from the sediment near the plant and into the water. A cleanup will likely involve dredging. The state has cleaned up 6000 sites, but 5000 remain on the list.

Over the last two or more centuries, humans have released large amounts of carbon dioxide into the atmosphere. A great deal of it has been absorbed by ocean waters, and as a result, the pH of the oceans has decreased by about 0.1 and will decrease by another 0.3–0.4 (Feely et al., 2012). Cold acidic water flows into the Sound. Shellfish (e.g., oysters, clams, mussels, crabs, abalone, and pteropods). The upwelling brings nutrients to the shellfish, but freshwater runoff also brings nitrogen and phosphorous into the Sound. Those nutrients encourage the growth of algae blooms that deplete oxygen levels. Hypoxia is a marker for these processes.

Moore et al. (2008) conducted an extensive survey of the characteristics of the Sound. They monitored 16 stations monthly from 1993 to 2002 and 40 stations biannually from 1998 to 2003. The temperature of the water is the warmest in September and the coolest in February. The salinity and density are greatly influenced by freshwater flowing in from the many rivers, especially in the winter and spring. The water varies according to many factors, including mixing over the sills between the basins, the depth of the water, distance from a river mouth, and river flows.

Freshwater

Rivers

Many rivers flow into the Puget Sound. The largest are the Cedar/Lake Washington canal, Duwamish/Green, Elwha, Nisqually, Nooksack, Puyallup, Skagit, Skokomish, Snohomish, and Stillaguamish Rivers, but thousands more rivers and streams also enter the Sound. Each year those combine to bring about 12 trillion gallons of fresh water into the Sound.

Those rivers, streams, and creeks transport about 6.5 million tons of sediment to the Sound every year (Czuba et al., 2011). Most of it (70%) comes from the rivers, but the rest (30%) is from shoreline erosion. The sediment consists of sand, silt, gravel, wood, and other material that is critical for the health of the Sound. The gravel and heavier matter settle out in the rivers. The lighter material is carried on into the Sound. Interestingly, the sediment is a blessing and a curse to the Sound. It maintains beaches and marshlands that are habitats for plants and animals, and without adequate replacement, habitat can be lost (Warrick et al., 2009). However, too much sediment can bury food and habitat (Grossman et al., 2011). Water quality is also affected since the sediments easily absorb toxins.

Seattle gets most of its water from two watersheds. The Cedar River Watershed drains 90,638 acres and provides drinking water for about 70% of Seattle's population. The Masonry Dam resulted in the Chester Morse Lake and the Masonry Pool reservoirs. The water also runs a hydroelectric power plant and is released to flow to the Landsburg diversion dam. There part of the water is sent to Lake Youngs Reservoir and on to the Cedar Water Treatment Facility. The Tolt River Watershed supplies about 30% of Seattle. Water from the Tolt Reservoir behind Tolt Dam flows into the South Fork Tolt River to the Tolt Water Treatment Facility.

Ground Water

The Puget Sound aquifer system underlies 7300 square miles (Vaccaro et al., 1998). About 117 square miles of that area are covered by surface water. The system can be divided into three main types. Alluvial and coarse-grained deposits form the aquifer units. Fine-grained deposits form confined or semiconfined units. Tertiary and older rock provides a confining basement for the aquifers. The annual recharge of the aquifers is, on average, 27 inches (0 to 70 inches range). The water is mostly of good quality. Seawater has intruded on some parts. Vaccaro et al. (1998) described the various aquifers that make up the whole system.

The Puget Sound system includes several aquifers. In Seattle, the Highline Well Field has three wells that produce ten million gallons each day (Gibson, nd). The field contains three aquifers. The upper aquifer is a shallow, unconfined acquired that is contained in sand. Unconfined means that the water is free to spread out through the matrix. The middle aquifer is confined in clay. The clay holds the water in the area of the aquifer. A deep aquifer is in sand and gravel with clay and silt also. Water is added to the aquifer through the same well. It flows under gravity in the space between the well casing and the pump column.

THE SOUND ITSELF

While the Puget Sound receives water inputs from many sources, the greatest is the Pacific Ocean. In recent years, the higher levels of carbon dioxide in the atmosphere due to climate change have caused a drop in the pH of the world's oceans, and the more acidic waters flow into the bottom of the estuaries and bays. How has the lower pH affected the Sound? Bianucci et al. (2018) examined that question. They studied the characteristics of carbonate-containing molecules and their effects on the Sound water. They concluded that both ocean and freshwater influence the Sound.

The Puget Sound region is a natural wonder with beautiful scenery and rich resources. However, it is also highly urbanized, and that development affects the natural environment and can erode its value. For example, coastal wetlands provide food, protect the coast, slow erosion, purify water, cycle nutrients, and maintain fisheries (Barbier et al., 2010). The Sound has traditionally been a productive region for oysters, clams, and mussels. A significant industry has developed to harvest these resources. Human activities are also associated with water-borne diseases from polluted water and contaminated seafood (Stewart et al., 2008).

The newer suburbs separate their sewer and stormwater systems, but Seattle is locked into an older system. Seattle and King Counties have a combined sewage system that carries both sewage and stormwater. Every year, more than 120 sewer overflow outfalls discharge about a billion gallons of untreated sewage, stormwater, and industrial wastewater into the Sound (Susewind, 2011). Microbial contamination has caused closure of some areas previously used for commercial and recreational shellfish harvesting.

Climate change will bring even greater challenges in urbanized coastal regions (Cutter et al., 2014). Increases in water pollution, warmer air, and water, and changes in precipitation will affect the natural environment. There are likely to be more extreme weather events, including floods, storms, and drought. Heavy rains will stress sewage systems, and drought will stress septic systems (Prudhomme et al., 2012). All of these will affect the quality of the water in the Sound.

POLLUTION

With its large population, the Puget Sound is quite polluted by many potentially toxic chemicals, especially in the sediments. The chemicals and heavy metals are found throughout the Sound, but are concentrated near the centers of population and industry, such as Elliot Bay, Commencement Bay, and Sinclair Inlet. More concerning is that these pollutants are concentrated in living organisms within the Sound (Hall, 2014).

In general, more lipophilic compounds have a greater tendency to accumulate in various organisms. Organic compounds are more readily accumulated in living things. Metals are much less lipid soluble but can still become concentrated in the gills of fish. However, if the metal becomes part of an organic compound, such as methyl mercury, it can then be concentrated in the fatty tissues. The highest levels of bioaccumulation occur at the top of the food chain because they have been concentrated at every step of the food chain.

Runoff from industrial operations and municipal wastewater and stormwater into the Puget Sound contains many pollutants, such as metals and organics [polychlorinated biphenyls (PCBs), polybrominated diphenyl ethers (PBDEs), polycyclic aromatic hydrocarbons (PAHs), and phthalates]. Many of these pollutants collect in the sediments at the bottom of the Sound. Microplastics are another significant source of pollution. They also accumulate other toxins. The Stormwater Work Group of Puget Sound has been monitoring pollution bound to sediments in the Sound (Black et al., 2018). They sampled 41 sites throughout the Sound. For most nearshore sediments, the pollutant levels were low and of no real concern. Microplastics were generally small fibers and again not in high concentrations.

Dealing with contaminated material is complicated (Palermo et al., 2000). The material could remain in place with no action, but that is usually not acceptable. Pipes can be used to move dredged material to diked areas nearshore or to upland confined disposal sites. The material can be left in an open water site and covered or capped with material that isolates the contaminated material. That material can also be moved to a landfill site for disposal. Palermo et al. (2000) summarized many of the federal and state requirements for the disposal of contaminated materials in the Puget Sound region.

The contaminants in the Sound include heavy metals, polycyclic aromatic hydrocarbons, polychlorinated hydrocarbons, phthalates, phenols, other organics, ammonia, cyanide, and hydrogen sulfide (LaLiberte and Ewing, 2006). Many do not break down easily and bioaccumulate. They are found in mussels, sole, rockfish, salmon, chinook salmon, seals, and killer whales. Some concentrate in sediments. Bioaccumulation occurs in mountain whitefish, bridgelip sucker, mussels, English sole, rockfish, chinook salmon, and other salmonids, seals, and killer whales.

SEWAGE IN PUGET SOUND

The Puget Sound is a large estuary system with multiple basins. It is dominated by waters from the Pacific Ocean: 70% of the dissolved nutrients come in through the Strait of Juan de Fuca. Still, local regions within the Sound have different environments with the influence of the many rivers and other sources of water and other material. The nutrient loads differ, and hypoxia occurs in some subbasins, and pollutants affect recreation and interests. The circulation patterns through the Puget Sound are fairly well understood. They are similar to classic fjords in many ways. Khangaonkar et al. (2011) updated the models in a more sophisticated manner. The model reproduces the velocity and salinity profile characteristics in the subbasins and circulation due to tides.

As development has increased, more people have moved from city centers into suburbs and the rural and undeveloped areas outside cities. The handling of sewage is a major problem in all of these areas. For example, the levels of nitrogen and phosphorus increase as more raw untreated sewage flow into the lakes and rivers. Those nutrients, in turn, encourage the growth of algae.

Moore et al. (2003) looked at three indicators of eutrophication in 30 lakes in the Seattle area. Those indicators were the concentrations of chlorophyll a and phosphorus and the algae that zooplankton do not eat. The lakes were divided into three groups: septic, sewer, and undeveloped lakes. Sewer lakes were connected to urban sewer systems. Septic lakes were on the urban-rural fringe. Mackas and Harrison (1997) found that eutrophication of the Puget Sound is unlikely due to two factors. First, the system has high levels of nitrate and ammonia already, and second, the tides turn over the water in the Sound fairly rapidly and thus wash out pollutants. They quantitated each of these and showed that locations inside the Sound and especially near discharge points vary in the amounts of pollutants. Areas in the Strait of Juan de Fuca had the least variation.

Nevertheless, fecal bacterial have been found at many locations in the Sound over the years. Those bacteria (e.g., *E. coli*, enterococci) are indicators of sewage contamination (Essington et al., 2011). In the Puget Sound, this contamination results from point-sources (e.g., sewer overflows) and non-point-sources (e.g., surface runoff). The Washington Departments of Health and Ecology monitor the Sound for these pollutants. Fecal coliform levels were also examined in 2001–2005 in the Sinclair-Dye Inlet near the Puget Sound Naval Shipyard at Bremerton (May et al., 2007). The goal was to quantify the levels and sources of the bacteria and to improve the water quality. They found multiple sources of bacteria and the paths of its transport throughout the Inlet. In general, their results implicated the typical sources of pollution. For example, areas with higher the population density had higher bacterial levels. Areas with older sewer infrastructure also had higher levels. Levels were higher after storms when runoff was greater, but these tended to be of short duration. More developed areas have more of the land surface covered with hard surfaces. They have more runoff too. Johnston et al. (2009) also modeled the coliform bacteria in these inlets.

Banas et al. (2015) sought to determine the water quality at the mouths of 15 major rivers that flow into the Sound. Using tracers released into the rivers, they established the circulation patterns that move nutrients and pollutants around the Sound throughout the year. They found that, in the spring, the fecal coliform bacteria and dissolved inorganic nutrients result from local sources. They also found that the Fraser River has a significant influence on the Sound. The most contaminated areas of the Sound in 2001–2005 were Budd Inlet, Commencement Bay, Oakland Bay, Port Angeles Harbor, Possession Sound, and Elliot Bay (Johnston et al., 2009). Other shellfish growing areas were also affected. Unfortunately, the levels have not varied much from 1998 to 2007.

Several sources have been implicated in the release of fecal bacteria into the Sound. One sourced involves septic tanks, which are widely used around the Sound. In 2012, there were 60,000 systems in the area. Those systems work well when properly cared for. Inventories of the systems have progressed well, but inspections are behind the targets. Also the Sound is a boater's dream and a major commercial port with 153,000 registered recreational vessels and 3600 commercial vessels (Jankowiak, 2018). Its long narrow shape makes it sensitive to sewage discharge from vessels. In 2017, the US EPA approved a plan to make the Sound a no-discharge area. Public hearings sponsored by Washington State University will result in the Sound becoming a no-discharge area. This is a major step forward in preserving the pristine nature of the Sound.

The bacteria can have other effects too. The presence of raw sewage in the waters of the Sound can also affect the state's production of farmed shellfish, a $270 million industry (Hamilton et al., 2016). The oysters and other shellfish absorb the bacteria, viruses, and other contaminants in the water and become unsafe for human consumption. The state has more than 100 commercial shellfish growing areas on 300,000 acres. These all require careful monitoring to ensure they are safe. Hormones are an often-overlooked pollutant. Johnson et al. (2008) examined the exposure of fish in the Puget Sound to estrogen. They assessed levels of vitellogenin, a yolk protein produced in mature females in response to estrogens. However, male fish also produce the protein when exposed to estrogen. Male sole (*Parophrys vetulus*) from several urban sites (e.g., Elliott Bay) had high levels of vitellogenin. The likely sources were industrial discharge, surface runoff, and sewer outfalls.

Not all of the effects of contaminants in the water are on the natural systems. They can also affect the perception of value in the property around the Sound. Papenfus (2019) looked at the effects of water quality impairments on the cost of housing in the region. He concluded that the higher levels of fecal coliform bacteria and dissolved oxygen lowered the value of houses in those areas. Coliform contaminants yielded the greatest effect.

CHEMICAL POLLUTANTS

Various organic compounds are common pollutants in Puget Sound and other bays and harbors. One biomonitoring system for contaminants is caged mussels. Takesue et al. (2019) tested for PAHs, polychlorinated biphenyls (PCBs), and other toxic elements by mussels from sediment. The mussels took up lead and possibly copper at all test sites. PAHs were absorbed at Liberty Bay, Eagle Harbor, and Smith Cove. PCBs were taken up at industrial sites. They concluded that uptake of toxins from sediments is one of several pathways.

Polycyclic Aromatic Hydrocarbons

PAHs are nearly ubiquitous contaminants of freshwater and marine sediments. They are produced when coal, oil, wood, garbage, or other materials are burned. So, some occur naturally, but most are the result of human activities. They are commonly generated in chemical production (e.g., naphthalene). Some are considered to be carcinogenic. As they are the products of burning, they tend to vary greatly in size and composition. Many are insoluble in water, but they do bind to particles and so can be contained in sediments in the water. They are typically found as complex mixtures of hundreds or even thousands of compounds, each with its own set of characteristics and risks. Neff et al. (2005) developed a method for determining the hazard of PAHs in sediments. They calculate a hazard quotient (HQ) for each PA and sum them to produce a hazard index. That measure can provide insight into identification of the causes of toxicity.

Most of the PAHs in the Puget Sound come from petroleum and combustion products, such as industrial discharge, creosote, runoff, and incineration and automobile emissions (Johnson, 2000). PAHs get into the water in spills of fuel oil, crude oil, and petroleum products, and from non-point sources. They are readily metabolized in most animals, and so, they do not accumulate. However, they and their metabolites can still be toxic.

PAHs affect living organisms. For example, Pacific herring (*Clupea pallasii*) spawn near the shore. West et al. (2014) determined the baseline of the PAHs in five spawning areas in the Puget Sound. They found that the amounts of PAHs in fish near the areas of the highest concentrations increased with development. Little to no PAHs were transferred from the mother to offspring. High levels of PAHs resulted in necrosis that was similar to that in experimental zebrafish.

Not all PAHs are of recent origin. Kuo et al. (2011) examined sediment cores to study the past 250 years for soot and graphitic black carbon and charcoal black carbon in the Sound and in Hood Canal. The Sound region has grown rapidly, but the Hood Canal is still sparsely populated. They attempted to reconstruct 250 years of deposits of two types of black carbon: charcoal and soot or graphitic black carbon. The peak amounts were in the 1940s for the Puget Sound and the 1970s for Hood Canal. They also concluded that more softwoods were burned in Washington than on the East Coast and that changes in climate also tracked well with the number of wildfires.

Polychlorinated Biphenyls

PCBs are oily clear or yellow liquids or solids. Because they are resistant to extreme temperatures and pressures, they were widely used in electrical equipment, hydraulic fluids, lubricants, plasticizers, and more. Although their manufacture was banned in 1979 by the US Environmental Protection Agency, they remain in the environment from earlier leaks, spills, and improper disposal. PCB levels are declining since no more are being made, but they are also persistent. PCBs collect in sediments and can accumulate in fish and shellfish.

PCBs leak from lamp ballasts, caulk, small, and large capacitors, and transformers and the amounts in everyday usage dwarf those lost into Puget Sound (200,000–500,000 versus 1500 kg). the greatest sources of PCBs are lamp ballasts (400–1500 kg) and industrial processes (900 kg). Davies and Delistraty (2016) recommended changes to prevent their loss into the environment. For example, replace their use where possible, reduce processes that generate them, better monitoring and education.

Pollution has wide-ranging effects on plants and animals. The classic example is the thinning of bird egg shells by dichlorodiphenyltrichloroethane (DDT). The thin shells break and allow the infant birds to die. The U.S. Fish and Wildlife Service sponsored a two-year study of the effects of pollutants on birds and fish at the South Bay Salt Works in the San Diego Bay National Wildlife Refuge (Zeeman et al., 2008). Failed egg shells of various species were examined for thickness and presence of pollutants, such as metals, organochlorine pesticides, organotins, PCB congeners, and polybrominated diphenyl ethers (PBDEs). The last two are related to DDT. The birds examined were black skimmers, Caspian terns, elegant terns, and least tern. The fish were California killifish, topsmelts, and longjaw mudsuckers. The egg shells were thinner than normal. The eggs had a high concentration of DDE and PCBs. The concentrations in the fish samples were much lower.

Forage fish in Puget Sound have high levels of toxic compounds that can accumulate as one goes up the food chain. These include a long list of organic compounds, such as PCBs, polybrominated diphenyl ethers, chlorinated pesticides, PAHs, alkylphenols, and chlorinated paraffins. Conn et al. (2020) found all of these in Pacific sand lance (*Ammodytes hexapterus*).

O'Neill and West (2009) studied the bioaccumulation of PCBs in Chinook salmon (*Oncorhynchus tshawytscha*) in the Puget Sound. They found that the PCB concentrations in fish from the Sound were three to five times greater than those in the same species from other points on the West Coast. Interestingly, most of the accumulated PCBs were gained in a marine environment, and age, size and lipids only accounted for 37% of the variation. Another factor was contributing to the excess bioaccumulation of Puget Sound salmon. The authors speculated that the additional PCBs were due to the residency in the Sound itself.

Harbor seals (*Phoca vitulina*) are among the top-level predators in the Puget Sound. As such, they accumulate contaminants to a greater degree than many other species. By examining them, much can be learned about multiple organic contaminants in the waters. Ross et al. (2004) biopsied 60 young harbor seals in Queen Charlotte Strait, Strait of Georgia, and the Puget Sound. They looked at levels of PCBs, polychlorinated dibenzo-p-dioxins (PCDDs), and polychlorinated dibenzofurans (PCDFs). Seals in the heavily industrialized Sound had higher concentrations of PCDDs and PCDFs.

Metals

Heavy metals come from a number of sources. For example, road dust often contains levels of copper lead, zinc, and platinum from vehicles (Hwang et al.,

2016). Lead is less common than in earlier years. The biggest reason is that gasoline no longer contains lead. However, other metals have increased in the environment with the greater numbers of vehicles in use. Runoff from rain carries the dust into rivers and streams and ultimately into the larger bodies, such as the Sound. The authors conclude that additional regulations would be helpful in controlling heavy metals in lakes, bays, and other large bodies of water.

Bloom and Crecelius (1987) surveyed for the presence of toxic heavy metals in the sediments of the Puget Sound. By examining sediment cores, they found moderate amounts of silver, mercury, lead, and copper but no cadmium. The metals were fairly well distributed around the Sound as well. The levels of metals were found by Brandenberger et al. (2008) to increase over the years. They examined sediment cores from sites near Seattle and Tacoma in 1982, 1991, and 2005. Levels of lead, copper, and arsenic increased over those years. They were also able to use their data to calculate the time it might take for those levels to return to pre-human levels. Arsenic has done that. However, while lead and copper might have recovered by 2020–2030, urbanization has increased the number of nonpoint sources in the 21st century, and those changes have extended the estimate of recovery to 2030–2060.

Mercury levels peaked during the Second World War. Sinclair Inlet, Bellingham Bay, and Elliott Bay all had very high levels (Paulson et al., 2010). Since then, mercury has been recognized as a serious contaminant, and new regulations have mitigated its release. As a result, mercury levels in the sediment have seen a considerable reduction. Amounts of mercury in various animals were normal to greater than normal.

Copper is a common contaminant and toxic to marine organisms, and its levels in estuaries are of interest. It comes from several sources. Copper leaches out of antifouling hull paint and is discharged from industrial plants and stormwater, and is released when sediments are resuspended (e.g., ship movements, dredging). Copper exists in seawater in multiple forms, including in dissolved complexes, colloids, and particles. However, the most toxic form is thought to be the free hydrated ion. In this form, it is chelated by a number of organic molecules. Copper is neurotoxic to fish and interferes with the peripheral olfactory nerves. McIntyre et al. (2008) examined the effects of copper on the olfactory epithelium of juvenile coho salmon (*Oncorhynchus kisutch*). The olfactory response was reduced in water with low ionic strength. It was not cured by increasing the pH or the ionic strength. However, it was improved as the amount of dissolved organic chemistry was increased.

Unknowns

The pollutants noted above, such as heavy metals, PAHS, and PCBs, are well known, but others remain mysteries. One example concerns the spawning of Pacific coho salmon. Migration of the coho salmon to their spawning grounds occurs in the fall when the rains begin and increase runoff into streams. Some surprises can even be found in areas that were thought to have been restored (Scholz et al., 2011). The fish began to display unusual behaviors and then died quickly. The females retained their eggs. The fish and water were tested for many contaminants, but none was identified as the culprit. The conclusion was that the fish in transition from salt to freshwater are sensitive to an unknown contaminant in runoff. Spromberg et al. (2016) showed that highway run-off was lethal to coho. They also showed that simple pre-treatment of filtering the runoff through soil columns removed the toxic materials and proposed using similar bioremediation treatments as a simple, cost-effective way to protect the salmon spawning habitat.

Plastics and Microplastics

Pollution does not always involve chemicals or metals. Sometimes it is more mundane, but no less deadly. Plastic has become a major problem for marine life. For hundreds of years, humans have used the oceans as a dumping ground, but in the 1950s, the problem has become much worse as plastics came into common use (Worm et al., 2017). Plastics take a very long time to degrade, if they ever do. They also tend to float so they can lurk almost invisible under water, and they are very hard to remove from the oceans.

Good et al. (2009) found that derelict fishing gear was killing large numbers of marine birds, and Good et al. (2010) examined the amount of that gear in Puget Sound. They found that, since 2002, hundreds of nets contained more than 32,000 marine animals. Most were invertebrates, but many were fish, birds and mammals. They conclude that the hazard is even greater than they documented and that it may be of great concern for conservation. Ghost fishing has become a worldwide problem (Lively and Good, 2018). Lost fishing gear continues to "fish." Traps, pots and nets are the main culprits. The gear remains a hazard for very long times. Gilardi et al. (2010) developed a model for predicting the mortality due to derelict gillnets. They also determined a cost:benefit ratio for removing the gear from the Puget Sound. The cost of removing the nets was a bargain compared to the cost in lost Dungeness crab during the lifetime of the net. Arthur et al. (2014) devised a strategy for dealing with this hazard. It included studies to determine the extent of fish mortality, assessing the economic effects on the fishing industry, working with the industry to find solutions, and developing policies as needed.

Microplastics are very small bits of plastics that are 0.3–5 mm. Some are made to be that size, but others result from the degradation of larger pieces of plastic (Betts, 2008). They result from many every-day products and processes, including personal care products, plastic products, fabrics and fishing line, photodegradation of plastic items, cigarette filters, and more. Worse still, many personal care products contain microbeads, and billions of these particles are released into river, streams, and bays every day in the US. For example, large amounts of microplastics are released during the mechanical and chemical actions of washing synthetic fabrics. De Falso et al. (2019) examined the effluent of household washing machines and found 124–308 mg per kg of fabric (640,000–1,500,000 particles). Microplastics are produced on land, but are washed into the Sound by run-off and water treatment plants. Water plants actually produce only small concentrations, but they also process so much water than the total number of microplastics released is huge. It can be millions of particles every day (Mason et al., 2016). Even a polystyrene cup lip can produce nearly 108 nanoparticles after 56 days in a weathering chamber (Lambert and Wagner, 2016). The particles are so small that they easily pass through the filters in wastewater treatment plants and eventually get into the food chain.

Microplastics are a serious problem in marine environments and estuaries (Masura et al., 2015). They can block the digestive tract. They remain in the environment for long times. Most importantly, they accumulate hazardous chemical and heavy metals. Brennecke et al. (2016) showed how heavy metals adhere to plastics. In harbor, such as those in the Puget Sound, antifouling paint is commonly used to protect boats. The seawater leaches copper and zinc from the paint, and those compounds are absorbed by polystyrene beads and polyvinyl chloride fragments in significant amounts. The levels of copper in some marinas can be high, and that has led some to suggest that copper-based antifouling paints should be phased out in favor of better alternatives (Carson et al., 2009).

Shellfish are a major industry in the Puget Sound. There have been some allegations that the plastic equipment used by the shellfish farmers is contributing to the microplastic contamination in the Sound. However, that equipment is highly regulated, and it is specially designed to resist degradation in saltwater and it is deployed underwater where it is protected from UV exposure. Therefore, Schoof and DeNike (2017) concluded that this gear is unlikely to contribute significantly to the microplastics in the Sound.

DREDGING AND FILLING

Dredging is important to maintain shipping channel, remove contaminants, and provide fill for development and wetland restoration. While all of these activities are beneficial, they are not without cost. Removal of sediments disrupts the ecology of the Sound, and even the placement of the spoils can be problematic. Bolam and Rees (2003) studied the impacts of dredging and found that macrofaunal communities that live in stressed environments are more resistant to changes than those in more stable environments. For example, invertebrates from stable environments took one to four years to recover from a dredging operation, whereas those from disturbed environments took only nine months. Thus, dredging and filling must be done with careful planning and study.

Dredging Shipping Lanes

The waters providing access to the ports in the Puget Sound need to be dredged periodically. For example, in early 2021, the Port of Tacoma proposed to dredge the ship berths in the port to 17 m below the mean lower low water level (SEPA, 2021). The project will move about 27,000 m^3 of material to Commencement Bay. As another example, the Skagit River brings 2.8 million tons of sediment into the Skagit Bay. Over time, that material accumulates and interferes with shipping. The US Army Corps of Engineers is considering a plan to dredge the channel. Michalsen and Brown (2019) reported on the feasibility of a new site for depositing dredged material in the Sound. As part of that study, they determined what happens to dredged material when it is dumped. They examined six potential sites. Three were eliminated easily, but the others seem to have potential.

Contaminated sediments need to be removed from the natural environment and be disposed of appropriately. Spoils that are considered unsuitable will be put into an approved landfill site. In 2015, the Environmental Protection Agency and the Port of Tacoma agree to clean up contaminated sediments in the Port (EPA, 2015). The Port wanted to expand its operations, but early on, it found tributyltin in the sediments. The agreement states that the Port will dredge about 40,000 m^3 of material and move it to an approved disposal site. The Puget Sound Naval Shipyard is the only facility on the West Coast that can dry dock the largest Navy ships (Channell, 2000). Over the years, the sediments at the shipyard have become contaminated with PCBs. In the surface sediments (0 to 3 ft), PCB concentrations are up to 80 mg/kg organic carbon. Also, the contaminants include mercury, arsenic, copper, lead, and PAHs. The options for mitigating the contamination include stabilization and solidification with binders, such as cement.

The spoils from dredging can also be useful for preparing areas for appropriate development or to maintain wetlands. The U.S. Army Corps of Engineers and the Environmental Protection Agency, and the Port of Everett, worked together to use the spoils from a long-term dredging project (EPA, 2007). They used the spoils to nourish beaches on Jetty Island. About 323,000 cubic yards of material was placed on the island with hydraulic dredging. The spoils helped to sustain the saltmarsh habitat. A few years later, another 239,000 cubic yards was used to restore the berm width and elevation. The success of this project is a model for other mitigation projects.

While dredging and filling are sometimes necessary and beneficial, both processes are complicated. While dredging is important, it affects plants and animals in the Sound. Penttila (2007) looked at the effects of dredging on forage fish. These small school fish are an important link in the marine food chain. Examples of forage fish include Pacific herring (*Clupea pallasi*), surf smelt (*Hypomesus pretiosus*), and Pacific sand lance (*Ammodytes hexapterus*), Northern anchovy (*Engraulis mordax*), Columbia River smelt (*Thaleichthys pacificus*), and longfin smelt (*Spirinchus thaleichthys*). Each has its own needs for feeding and spawning, and the dredging and other human activities.

Dredging is also used to remove contaminants from sediments in the Puget Sound. However, the disposal of the spoils of dredging is a complex environmental problem. The US EPA estimates that about 10% of the sediments around the US are contaminated to levels that pose a risk to animals and humans (Leschine et al., 2003). However, there is no governmental mechanism exists to remedy the problem. In 1991, Washington state took it on themselves to develop management standards. They devised the Advocacy Coalition Framework to assess risk and hazards, and that tool has provided a way to find a path forward on sticky ecological clean-up questions. More recently, Sotirov and Memmler

(2012) evaluated the Framework and suggested some tweaks to it that might make it even more useful.

Evaluating Operations

How is one to judge the success of a recovery after a dredging or dumping operation (Wilber and Clarke, 2007)? Typically, assessments involve surveys of invertebrates and a number of physical factors. Many factors are involved, and researchers differ on the answer. Even the definition of term "recovery" is in dispute. Some belief the area should be judged for whether it returns to its pre-dredging state or should it be just as good as a similar reference area. In any case, finding a suitable reference area is difficult.

A demonstration project was conducted at the Shipyard. The system included activated carbon and an aggregate and was used to treat the sediments to remove various contaminants (Johnston et al., 2014). For several years, the site has been surveyed for leaching. Those surveys showed that the mix had achieved the performance standards that had been set.

BETWEEN LAND AND WATER

Wetlands are valuable areas for mitigating floods, storing carbon, and absorbing contaminants (Encyclopedia, 2012b). They provide habitats for many species and nurseries for others. The Puget Sound features a diverse ecosystem with enormous numbers of waterfowl, shorebirds, and raptors, shellfish, marine mammals, large runs of Pacific salmon, and many more plants and animals (Fresh et al., 2011). With climate change and sea-level rise, they offer a buffer to hold back at least part of the damage. Unfortunately, the value of wetlands was not recognized for many years, and so, much of the wetlands have been lost to dredging, filling, and development. Thus, the original shorelines in many areas are long gone.

This is certainly the case in the Puget Sound. In the last couple of centuries, multiple changes have come to the Sound. River deltas have lost area and shoreline. About 95% of the embayments were developed. Beaches and bluffs have been armored. About 27% of the shoreline of Puget Sound is armored. Much of the tidal wetlands have been lost. The shoreline is 15% less than what it was. Many places have multiple changes: 40% of the shoreline has been altered (e.g., structures, roads, and marinas). Collins and Sheikh (2005) found only 5650 hectares of wetlands in the Sound region, and about half of those are in the Skagit, Stillaguamish, and Samish river deltas. The median size was just 0.57 hectares. These numbers show that the tidal wetlands are now 17–19% of what they were.

The supratidal zone is that area between the purely marine and purely terrestrial ecosystems. It is an important area, and also one that is often disturbed by humans and armoring. Using paired sampling, Sobocinski et al. (2010) found that natural beaches had more oligochetes, nematodes, talitrid amphipods, insects, and collembolans than armored beaches. Armored beaches had crustaceans.

Much of the Puget Sound shoreline is armored, and those structures affect the native environment. Morley et al. (2012) mapped the armoring on the Duwamish River estuary and documented its various characteristics (e.g., temperature, types of invertebrates, salmon diet, degree of armoring). The water was the same at armored and unarmored sites. Invertebrates were more diverse at unarmored sites.

The changes have dramatically affected the native species. For example, the production of many species of fish, shellfish, and wildlife has been impaired. Several salmon species are now protected by the federal Endangered Species Act. Forage fish spawning areas have been lost. Shellfish beds are threatened by contamination. Bird numbers have declined. As a result, shore environments are at risk of further degradation from climate change and continued human development in the region.

One example of the Puget Sound wetlands is the Duwamish estuary. Juvenile Chinook salmon use the estuary as a hatchery, and recent projects have attempted to restore the estuary. Cordell et al. (2011) examined the results of those efforts. They found that the salmon in the restored areas fed on different species than in a natural environment and that the growth of the fish was greater at the natural site.

Five species of Pacific salmon spawn and rear in Puget Sound (Fresh, 2006). They complete their transition from freshwater to saltwater in the wetlands of the Sound. The different species spend different amounts of time in the tidal areas. For example, juvenile Chinook salmon fry enter natal deltas between December and April. Some of the spend only days in deltas. Others stay for up to 120 days. In late spring, parr migrants and yearlings arrive in the estuary. Fish size at the time the fish arrive in the delta and residence time in the delta tend to be inversely related. Water temperature may also be important for when juvenile Chinook salmon leave the delta. Once they leave the delta for the Puget Sound, normally in June and July, they distribute widely throughout nearshore ecosystems. Juvenile Chinook use deeper and offshore habitats, and in the end, most leave to feed in the North Pacific Ocean.

CLIMATE AND AIR QUALITY

CLIMATE

The Puget Sound has a reputation for rain, but that is not completely deserved. Summers are dry, but from late September to June, there is rain, and it is not always heavy. In the Sound, geography determines the weather. The mountains of the Olympic Peninsula reach nearly 2400 m. With that height, they cast a "rain shadow" over much of the Sound. The rain falls on the western side of the Olympics before it gets across the mountains. This leaves some areas of the Sound quite dry even in the "rainy" season.

The weather in the Northwest is governed by much larger forces. In mid-October, the Aleutian low-pressure cell moves into the Gulf of Alaska (Encyclopedia, 2012a). Its surface winds blow counterclockwise. Off Southern

California, a high-pressure cell forms to circulate its winds in a clockwise manner. The combination of these two cells feeds moist air into the Pacific Northwest, which then falls as rain or snow as it rises up to cross the Olympic or Cascade mountains. In the spring, the Aleutian Low retreats northward and the high pressure moves north. Those movements end the rain in the Puget Sound. Variations in this pattern depend on the Aleutian Low. When it is weak, the Sound has a warm and dry winter. When it is strong, the weather is cooler and wetter. The Low varies with two factors: the El Nino-Southern Oscillation and the Pacific Decadal Oscillation.

AIR QUALITY

Although Seattle has fairly good air quality, there is room for improvement (IQAir, 2021). The US Environmental Protection Agency recommends no more than three unhealthy days per year for each pollutant. Unfortunately, in recent years, Seattle has had about 14 unhealthy days per year. Seven of those are due to high levels of ozone, and seven more are due to particulate matter. When US metropolitan areas are ranked for air quality, Seattle comes in at 36 of 229 for worst pollution due to ozone and 14 of 216 for particulate matter. Most of Seattle's pollution comes from vehicle traffic. Industrial facilities and ship traffic contribute to the pollution, but heavy trucks are still the main source of particulate matter in the air in Seattle. Other significant sources are wildfires and the burning of wood in fireplaces. More concerning is that the recent trend for rising levels of those pollutants.

Ozone and particulate matter both irritate the respiratory system and can be especially harmful to those with asthma. Snow et al. (2003) examined levels of these pollutants in the air over Tacoma. They may three flights over the Sound to collect data on the pollutants and other factors. The flights showed that levels of ozone and particulate matter were enhanced and reached 4 km over the Puget Sound in the summer. They also showed the value of airplane data collection to track pollution levels.

Although one might assume that air pollution affects everyone the same since it is in the air, this is not the case. The concentrations of pollutants sometimes do not spread very far. For example, homes near freeways are often less expensive, and the residents there are exposed to pollutants that have their highest levels within 100 m of the freeway. Bae et al. (2007) documented high levels of pollutants close to roadways in Seattle and Portland.

Interestingly, some of the contaminants in the water of the Puget Sound begin as air pollution (Dunagan, 2020). Airborne pollutants fall to the ground and are washed into the Sound by stormwater. For example, wildfires send lots of pollutants into the air. Fish in seemingly pristine lakes high in the mountains of Olympic National Park have unhealthy levels of lead. Other air pollutants that can find their way to water include polychlorinated biphenyls, polybrominated diphenyl ethers, and polycyclic aromatic hydrocarbons.

Atmospheric Rivers

In mid-January 2021, the Pacific Northwest was hit by several heavy rainstorms (Hansen, 2021). Some areas of the Olympic Peninsula received 10 inches of rain. The storms were part of an atmospheric river. Atmospheric rivers are narrow streams of water vapor thousands of miles long. They can reach from the Hawaiian Islands to the West Coast of the United States, and they bring enormous amounts of rain and snow to the western mountains. The amount of water they carry is hard to imagine. One atmospheric river can carry water that would equal 7–15 of the Mississippi River (Ralph and Dettinger, 2011).

Neiman et al. (2011) studied the factors involved in flooding in four major watersheds in Western Washington. They found that the common factor was atmospheric rivers. Although some factors were different, the flooding occurred when the precipitation was mostly rain. Over the last 40 years, 80% of the flood damage on the West Coast resulted from atmospheric rivers. Yet, the massive amounts of rain are not always detrimental. They bring 30–50% of the precipitation to the West Coast, and they have tended to end droughts on the West Coast (Dettinger, 2013). In the Pacific Northwest, 60–74% of droughts have ended with an atmospheric river.

Corringham et al. (2019) used a scale to rank atmospheric rivers on a scale from 1 to 5 that is approximately exponential. They found the extent of the damages increases exponentially with the storm intensity and duration. Storms in the future will likely be even more dangerous. The population will increase and so will the infrastructure to support them. Finally, climate change will be a major influence on the storms.

Wildfires

The number of wildfires in the Western United States and Washington has increased in recent decades, and global warming is likely to accelerate wildfires. For example, September 2020 was an unusually bad month for wildfires in Washington. Fires burned 330,000 acres in 24 hours. Most of the burning was in the eastern part of the state, but the Puget Sound region had its share of wildfires. The region is already experiencing reduced precipitation (Abatzoglou et al., 2014) and higher temperatures. Furthermore, the incidence of wildfires in the Puget Sound region is likely to increase as global temperatures rise (Barbero et al., 2015). Abatzoglou and Williams (2016) showed that human activities caused half of the increase in dry fuel since the 1970s. They believe that trend is unlikely to end. Land use planning and fire abatement measures can help in many cases (Syphard et al., 2013), but those efforts are expensive and often is in conflict with the desire of many to enjoy the benefits of living close to nature. The climate crisis, with its higher temperatures and stronger winds, is likely to increase the number and intensity of wildfires and the threat to humans and their property.

Many forests in the Puget Sound region have been stable for decades. These old-stand forests can withstand stressful conditions for much longer times than forests that have turned over multiple times. Halofsky et al. (2018) called this phenomenon "landscape inertia." They modeled the effects of climate change and fire on the forests in western Washington through the end of the 21st century. They found that percentages of some species may change. For example, Douglas fir tolerates fire and drought better than western hemlock and western red cedar, and so, the firs may become the dominant species at lower elevations, in the Puget lowlands and northeastern side of the Olympic Mountains. This was also the case during drier times in prehistoric forests. Also, species will expand or contract their boundaries. Western hemlock will extend their range farther up the mountains.

Higher temperatures and more fires are not a new thing. To study the use of fossil fuels and biomass, Kuo et al. (2011) looked at the deposition of black carbon in the Sound and in Hood Canal. The Sound region has grown rapidly, but the Hood Canal is still sparsely populated. They attempted to reconstruct 250 years of deposits of two types of black carbon: charcoal and soot/graphitic black carbon. The peak amounts were in the 1940s for the Puget Sound and the 1970s for Hood Canal. They also concluded that more softwoods were burned in Washington than on the East Coast and that changes in climate also tracked well with the number of wildfires.

Smoke from wildfires has become a hazard in the region. In 2018, wildfire smoke resulted in 24 days of poor air quality. Nine of those days were unhealthy for sensitive groups or unhealthy for everyone. Zauscher et al. (2013) examined the effect of fires on the local air quality. They found that 84% of the 120–400-nm particles were biomass burning aerosols. Those particles absorb solar radiation and serve as nuclei for cloud condensation. They also are associated with adverse health effects. The make-up of the particles also varies according to the temperature of the fire. Hotter fires produce particles with more inorganic material and more soot. Cooler smoldering fires produce more organic carbon-rich aerosols. Lassman et al. (2017) studied the methods of measuring the exposure of people to smoke, a more difficult task than might be assumed due to the rapid changes in smoke concentrations Gan et al. (2017) examined health outcomes and population exposure to smoke from wildfires in Washington 2012. They compared hospital admissions sorted by ZIP code to smoke levels determined by different methods. The association of smoke and health effects depended on the method used.

Littell et al. (2010) noted that climate change will make the conditions for wildfires much worse. Temperatures will be higher and summers will be drier. They showed that July and August temperature and precipitation or the July and August water-balance deficit can account for about half of the variability in the area burned by wildfires. Their projections for the future are frightening. The area burned is likely to double or triple by the 2080s.

CONCLUSIONS

The Puget Sound was created by volcanoes, glaciers, waves, water, and plate movements over millions of years. The last glaciers retreated about 12,000 years ago and released their water. Since then, the Sound region has been relatively stable. Then humans arrived. Although people have only been in the region for a short time, right now, they are the dominant force for change in the Puget Sound region. Sadly, some of that chance has been at the expense of the natural environment. The Native Americans were better stewards than those of us who arrived later. Yet even they are implicated in the extinction of large mammals, and they certainly used fire to clear land. However, the Europeans or Americans have made the most changes. The region is heavily populated and developed. Much of the land has been paved and built up. Rivers have been dammed and channeled. The Sound has become polluted. We may be entering a sixth mass extinction that is caused by humans, and this one might threaten even our own species. Fortunately, there is a growing realization of the importance of the environment. That is detailed in Chapter 9.

REFERENCES

Abatzoglou JT, Rupp DE, Mote PW (2014) Seasonal climate variability and change in the Pacific Northwest of the United States. *Journal of Climate* 27(5): 2125–2142.

Abatzoglou JT, Williams AP (2016) Impact of anthropogenic climate change on wildfire across western US forests. *Proceedings of the National Academy of Sciences of the USA* 113: 11770–11775.

Arthur C, Sutton-Grier AE, Murphy P, Bamford H (2014) Out of sight but not out of mind: Harmful effects of derelict traps in selected U.S. coastal waters. *Marine Pollution Bulletin* 86: 19–28.

Bae CHC, Sandlin G, Bassok A, Kim S (2007) The exposure of disadvantaged populations in freeway air-pollution sheds: A case study of the Seattle and Portland regions. *Environment and Planning B: Planning and Design* 34: 154–170.

Balk G (2016) *Seattle Among Top 10 Most Densely Populated Big Cities in the U.S. for First Time Ever*. Seattle Times. Retrieved from: https://www.seattletimes.com/seattle-news/data/seattle-density-doesnt-have-to-be-a-dirty-word/. April 21, 2021.

Balk G (2019) *The Decade in Demographics: Top 5 Changes in the Seattle Area*. Seattle Times. Retrieved from: https://www.seattletimes.com/seattle-news/data/the-decade-in-demographics-seattles-top-5-changes/#:~:text=Seattle's%20population%20has%20ballooned%20by,only%20grew%20by%20about%20116%2C000. April 21, 2021.

Banas NS, Conway-Cranos L, Sutherland DA, MacCready P, Kiffney P, Plummer M (2015) Patterns of river influence and connectivity among subbasins of Puget Sound, with application to bacterial and nutrient loading. *Estuaries and Coasts* 38: 735–753.

Barbero R, Abatzoglou JT, Larkin NK, Kolden CA, Stocks B (2015) Climate change presents increased potential for very large fires in the contiguous United States. *International Journal of Wildland Fire* 24: 892.

Barbier EB, Hacker SD, Kennedy C, Koch EW, Stier AC, Silliman BR (2010) The value of estuarine and coastal ecosystem services. *Ecological Monographs* 81(2): 169–193. DOI: 10.1890/10-1510.1.

Baum RL, Coe JA, Godt JW, Harp EL, Reid ME, Savage WZ, Schulz WH, Brien DL, Chleborad AF, McKenna JP, Michael JA (2005) Regional landslide-hazard assessment for Seattle, Washington, USA. *Landslides* 2: 266–279.

Betts K (2008) Why small plastic particles may pose a big problem in the oceans. *Environmental Science & Technology* 42: 8995.

Bianucci L, Long W, Khangaonkar T, Pelletier G, Ahmed A, Mohamedali T, Roberts M, Figueroa-Kaminsky C (2018) Sensitivity of the regional ocean acidification and carbonate system in Puget Sound to ocean and freshwater inputs. *Elementa: Science of the Anthropocene* 6: 22. DOI: 10.1525/elementa.151.

Black RW, Barnes A, Elliot C, Lanksbury J (2018) *Nearshore Sediment Monitoring for the Stormwater Action Monitoring (SAM) Program, Puget Sound, Western Washington.* U.S. Geological Survey, Washington. p. 53. DOI: 10.3133/sir20185076.

Bloom NS, Crecelius EA (1987) Distribution of silver, mercury, lead, copper and cadmium in Central Puget Sound sediments. *Marine Chemistry* 21: 377–390.

Bolam SG, Rees HL (2003) Minimizing impacts of maintenance dredged material disposal in the coastal environment: A habitat approach. *Environmental Management* 32: 171–188.

Brandenberger JM, Crecelius EA, Louchouarn P (2008) Historical inputs and natural recovery rates for heavy metals and organic biomarkers in Puget Sound during the 20th Century. *Environmental Science & Technology* 42: 6786–6790.

Brennecke D, Duarte B, Paiva F, Caçador I, Canning-Clode J (2016) Microplastics as vector for heavy metal contamination from the marine environment. *Estuarine, Coastal and Shelf Science* 178: 189–195.

Carson RT, Damon M, Gonzalez JA, Johnson LT (2009) Conceptual issues in designing a policy to phase out metal-based antifouling paints on recreational boats. *Journal of Environmental Management* 90: 2460–2468.

Channell MG (2000) *An Evaluation of Solidification/Stabilization for Sediments from the Puget Sound Naval Shipyard.* U.S. Army Engineer Research and Development Center, Vicksburg, MS.

Coe JA, Michael JA, Crovelli RA, Savage WZ, Laprade WT, Nashem WD (2004) Probabilistic assessment of precipitation triggered landslides using historical records of landslide occurrence, Seattle, Washington. *Environmental and Engineering Geoscience* X(2): 103–122.

Collins BD, AJ Sheikh (2005) *Historical Reconstruction, Classification, and Change Analysis of Puget Sound Tidal Marshes.* University of Washington (Seattle, WA) and the Nearshore Habitat Program, Washington State Dept. of Natural Resources, Olympia, WA.

Conn KE, Liedtke TL, Takesue RK, Dinicola RS (2020) Legacy and current-use toxic contaminants in Pacific sand lance (*Ammodytes personatus*) from Puget Sound, Washington, USA. *Marine Pollution Bulletin* 158: 111287.

Cordell JR, Ayesh DT, Ruggerone GT, Cooksey M (2011) Functions of restored wetlands for juvenile salmon in an industrialized estuary. *Ecological Engineering* 37: 343–353.

Corringham TW, Ralph FM, Gershunov A, Cayan DR, Talbot CA (2019) Atmospheric rivers drive flood damages in the western United States. *Science Advances* 5: eaax4631.

Cutter SL, Solecki W, Bragado N, Carmin J, Fragkias M, Ruth M, Wilbanks T (2014) Urban systems, infrastructure, and vulnerability. In: *Climate Change Impacts in the United States: The Third National Climate Assessment,* pp. 282–296. U.S. Global Change Research Program. Retrieved from: http://nca2014.globalchange.gov/report. April 21, 2021.

Czuba JA, Magirl CS, Czuba CR, Grossman EE, Curran CA, Gendaszek AS, Dinicola RS (2011) *Sediment Load From Major Rivers into Puget Sound and Its Adjacent Waters*. U.S. Geological Survey, Washington. pp. 2011–3083. Retrieved from: https://pubs.usgs.gov/fs/2011/3083/. April 22, 2021.

Davies H, Delistraty D (2016) Evaluation of PCB sources and releases for identifying priorities to reduce PCBs in Washington state (USA). *Environmental Science and Pollution Research International* 23: 2033–2041.

De Falso F, Di Pace E, Cocca M, Avella M (2019) The contribution of washing processes of synthetic clothes to microplastic pollution. *Scientific Reports* 9: 6633.

De Freytas-Tamura K (2018) *Plastics Pile Up as China Refuses to Take the West's Recycling*. New York Times. Retrieved from: https://www.nytimes.com/2018/01/11/world/china-recyclables-ban.html. April 21, 2021.

Dettinger MD (2013) Atmospheric rivers as drought busters on the U.S. West Coast. *Journal of Hydrometeorology* 14: 1721–1732.

Dunagan C (2020) *How Air Pollution Becomes Water Pollution with Long-Term Effects on Puget Sound*. Puget Sound Institute. University of Washington. Retrieved from: https://www.pugetsoundinstitute.org/2020/01/how-air-pollution-becomes-water-pollution-with-long-term-effects-on-puget-sound/. April 24, 2021.

Ecology (nd) *Puget Sound Feeder Bluffs*. Department of Ecology, Washington State. Retrieved from: https://ecology.wa.gov/Research-Data/Monitoring-assessment/Coastal-monitoring-assessment/Projects/Puget-Sound-feeder-bluff. April 16, 2021.

Encyclopedia (2012a) *Puget Sound's Climate*. Encyclopedia of Puget Sound. Retrieved from: https://www.eopugetsound.org/articles/puget-sounds-climate. April 24, 2021.

Encyclopedia (2012b) *Tidal Wetlands*. Encyclopedia of Puget Sound. Retrieved from: https://www.eopugetsound.org/science-review/3-tidal-wetlands. April 23, 2021.

EPA (2007) *Case Study: Jetty Island, Puget Sound*. US Environmental Protection Agency. Retrieved from: https://www.epa.gov/sites/production/files/2015-09/documents/beneficial-use-of-dredged-materials-case-study-of-jetty-island-puget-sound.pdf. April 17, 2021.

EPA (2015) *Case Summary: Cleanup Settlement Addresses Contaminated Sediment Removal in Blair Waterway, Washington State*. Environmental Protection Agency. Retrieved from: https://www.epa.gov/enforcement/case-summary-cleanup-settlement-addresses-contaminated-sediment-removal-blair-waterway. April 28, 2021.

Essington T, Klinger T, Conway-Cranos T, Buchanan J, James A, Kershner J, Logan I, West J (2011) *Marine Fecal Bacteria*. Encyclopedia of Puget Sound. Retrieved from: https://www.eopugetsound.org/articles/marine-fecal-bacteria. April 19, 2021.

Feely RA, Klinger T, Newton JA, Chadsey M (2012) *Scientific Summary of Ocean Acidification in Washington State Marine Waters*. NOAA OAR Special Report #3934.

Fresh KL (2006) *Juvenile Pacific Salmon and the Nearshore Ecosystem of Puget Sound*. Puget Sound Nearshore Partnership Report No. 2006-06. Seattle District, U.S. Army Corps of Engineers, Seattle, Washington.

Fresh K, Dethier M, Simenstad C, Logsdon M, Shipman H, Tanner C, Leschine T, Mumford T, Gelfenbaum G, Shuman R, Newton J (2011) *Implications of Observed Anthropogenic Changes to the Nearshore Ecosystems in Puget Sound*. Puget Sound Nearshore Ecosystem Restoration Project Technical Report 2011-03.

Gan RW, Ford B, Lassman L, Pfister G, Vaidyanathan A, Fischer E, Volckens J, Pierce JR, Magzamen S (2017) Comparison of wildfire smoke estimation methods and associations with cardiopulmonary-related hospital admissions. *GeoHealth* 1: 122–136.

Gibson M (nd) *Seattle*. Groundwater Geek. Retrieved from: http://www.groundwatergeek.com/asr-by-state/washington/seattle. April 23, 2021.

Gilardi KVK, Carlson-Bremer DC, June JA, Antonelis K, Broadhurst G, Cowan T (2010) Marine species mortality in derelict fishing nets in Puget Sound, WA and the cost/benefits of derelict net removal. *Marine Pollution Bulletin* 60: 376–382.

Godt JW, Baum RL, Chleborad AF (2006) Rainfall characteristics for shallow landsliding in Seattle, Washington, USA. *Earth Surface Processes and Landforms* 31: 97–110.

Gomberg J, Sherrod B, Weaver C, Frankel A (2010) *A Magnitude 7.1 Earthquake in the Tacoma Fault Zone: A Plausible Scenario for the Southern Puget Sound Region, Washington*. U.S. Geological Survey, Washington. p. 4.

Good TP, June JA, Etnier MA, Broadhurst G (2009) Ghosts of the Salish Sea: Threats to marine birds in Puget Sound and the Northwest Straits from derelict fishing gear. *Marine Ornithology* 37: 67–76.

Good TP, June JA, Etnier MA, Broadhurst G (2010) Derelict fishing nets in Puget Sound and the Northwest Straits: Patterns and threats to marine fauna. *Marine Pollution Bulletin* 60: 39–50.

Grossman, EE, George, DA, Lam, A (2011) *Shallow Stratigraphy of the Skagit River Delta, Washington, USA Derived from Sediment Cores*. U.S. Geological Survey, Washington.

Hall JE (2014) Bioconcentration, bioaccumulation, and the biomagnification in Puget sound biota: Assessing the ecological risk of chemical contaminants in Puget sound. *Tahoma West Literary Arts Magazine* 8: 40–51.

Halofsky JS, Conklin DR, Donato DC, Halofsky JE, Kim JB (2018) Climate change, wildfire, and vegetation shifts in a high-inertia forest landscape: Western Washington, U.S.A. *PLoS ONE* 13(12): e0209490.

Hamilton LJ, Berbells S, Sullivan L, Snyder J (2016) *Fecal Coliform Indicator Trends in the Puget Sound: Rain or Restoration?* Salish Sea Ecosystem Conference. Retrieved from: https://cedar.wwu.edu/ssec/2016ssec/policy_and_management/2/. April 19, 2021.

Hansen K (2021) *Potent Atmospheric Rivers Douse the Pacific Northwest*. Earth Observatory. Retrieved from: https://earthobservatory.nasa.gov/images/147822/potent-atmospheric-rivers-douse-the-pacific-northwest. April 24, 2021.

Hungr O, Leroueil S, Picarelli L (2014) The Varnes classification of landslide types, an update. *Landslides* 11: 167–194.

Hwang H-M, Fiala MJ, Park D, Wade TL (2016) Review of pollutants in urban road dust and stormwater runoff: part 1. Heavy metals released from vehicles. *International Journal of Urban Sciences* 5934: 1–27. DOI: 10.1080/12265934.2016.1193041.

Hyndman RD, Mazzotti S, Weichert D, Rogers GC (2003) Frequency of large crustal earthquakes in Puget Sound – Southern Georgia Strait predicted from geodetic and geological deformation rates. *Journal of Geophysical Research* 108(B1): 2033.

IQAir (2021) *Air Quality in Seattle*. IQAir. Retrieved from: https://www.iqair.com/us/usa/washington/seattle. April 24, 2021.

Jankowiak A (2018) *Puget Sound No Discharge Zone for Vessel Sewage*. Salish Sea Ecosystem Conference. Retrieved from: https://cedar.wwu.edu/ssec/2018ssec/allsessions/373/. April 19, 2021.

Johannessen JW (2001) *Soft Shore Protection as an Alternative to Bulkheads—Projects and Monitoring*. Puget Sound Research Station, Washington, D.C. p. 13.

Johannessen J, MacLennan A (2007) *Beaches and Bluffs of Puget Sound*. Puget Sound Nearshore Partnership Report No. 2007-04. Seattle District, U.S. Army Corps of Engineers, Seattle, Washington. Retrieved from: https://tamug-ir.tdl.org/bitstream/handle/1969.3/28805/beaches_bluffs[1].pdf?sequence=1. April 15, 2021.

Johnson LL, Lomax DP, Myers MS, Olson OP, Sol SY, O'Neill SM, West J, Collier TK (2008) Xenoestrogen exposure and effects in English sole (*Parophrys vetulus*) from Puget Sound, WA. *Aquatic Toxicology* 88: 29–38.

Johnson L (2000) *An Analysis in Support of Sediment Quality Thresholds for Polycyclic Aromatic Hydrocarbons (PAHs) to Protect Estuarine Fish.* US National Oceanic and Atmospheric Administration, National Marine Fisheries Service, Seattle, WA.

Johnston R, Chadwick B, Kirtay V, Rosen G, Guerrero J, Germano J, Conder J, Stransky C (2014) *Monitoring to Assess the Effectiveness of an Activated Carbon Sediment Amendment to Remediate Contamination at a Site Located at the Puget Sound Naval Shipyard & Intermediate Maintenance Facility, Bremerton, WA.* Salish Sea Ecosystem Conference. Retrieved from: https://cedar.wwu.edu/ssec/2014ssec/Day2/80/. April 17, 2021.

Johnston RK, Wang PF, Loy EC, Blake AC, Richter KE, Brand MC, Kyburg CE, Skahill BE, May CW, Cullinan MV, Choi W, Whitney VS, Leisle DE, Beckwith LB (2009) *An Integrated Watershed and Receiving Water Model for Fecal Coliform Fate and Transport in Sinclair and Dyes Inlets, Puget Sound, WA.* Space and Naval Warfare Systems Center, Technical Report 1977. Retrieved from: https://apps.dtic.mil/sti/pdfs/ADA536960.pdf. April 19, 2021.

Khangaonkar T, Yang Z, Kim T, Roberts M (2011) Tidally averaged circulation in Puget Sound sub-basins: Comparison of historical data, analytical model, and numerical model. *Estuarine, Coastal and Shelf Science* 93: 305–319.

King (2020) *Landfills.* King County. Retrieved from: https://kingcounty.gov/depts/dnrp/solid-waste/facilities/landfills.aspx. April 21, 2021.

Kuo L-J, Louchouarn P, Herbert BE, Brandenberger JM, Wade TL, Crecelius E (2011) Combustion-derived substances in deep basins of Puget Sound: Historical inputs from fossil fuel and biomass combustion. *Environmental Pollution* 159: 983–990.

LaLiberte D, Ewing RD (2006) *Effect on Puget Sound Chinook Salmon of NPDES Authorized Toxic Discharges as Permitted by Washington Department of Ecology.* Washington Department of Ecology. Retrieved from: https://www.peer.org/wp-content/uploads/attachments/06_19_6_report.pdf. April 24, 2021.

Lambert S, Wagner M (2016) Characterisation of nanoplastics during the degradation of polystyrene. *Chemosphere* 145: 265–268.

Lassman W, Ford B, Gan RW, Pfister G, Magzamen S, Fischer EV, Pierce JR (2017) Spatial and temporal estimates of population exposure to wildfire smoke during the Washington state 2012 wildfire season using blended model, satellite, and in situ data. *GeoHealth* 1: 106–121.

Leschine TM, Lind KA, Sharma R (2003) Beliefs, values and technical assessment in environmental management: Contaminated sediments in Puget Sound. *Coastal Management* 31: 1–24.

Littell JS, Oneil EE, McKenzie D, Hicke JA, Lutz JA, Norheim RA, Elsner MM (2010) Forest ecosystems, disturbance, and climatic change in Washington State, USA. *Clim Change* 102: 129–158.

Lively JA, Good TP (2018) Ghost fishing. In: *World Seas: An Environmental Evaluation Volume III: Ecological Issues and Environmental Impacts*, Academic Press, Cambridge, MA, pp. 183–196. DOI: 10.1016/B978-0-12-805052-1.00010-3.

Mackas DL, Harrison PJ (1997) Nitrogenous nutrient sources and sinks in the Juan de Fuca Strait/Strait of Georgia/Puget Sound Estuarine System: Assessing the potential for eutrophication. *Estuarine, Coastal and Shelf Science* 44: 1–21.

Mason SA, Garneau D, Sutton R, Chu Y, Ehmann K, Barnes J, Fink P, Papazissimos D, Roger DL (2016) Microplastic pollution is widely detected in US municipal wastewater treatment plant effluent. *Environmental Pollution* 218: 1045–1054.

Masura J, Baker J, Foster G, Arthur C, Herring C (2015) *Laboratory Methods for the Analysis of Microplastics in the Marine Environment: Recommendations for Quantifying Synthetic Particles in Waters and Sediments.* NOAA Technical Memorandum NOS-OR&R-48.

May C, Cullinan V, Johnston R, Wang PF, Sherrell G, Leisle D, Skahill B, Lawrence S, Roberts M, Keiss J, Ostrom T (2007) An assessment of fecal coliform bacteria in the Sinclair/Dyes Inlet Watershed, Puget Sound, WA, USA. Proceedings of the Georgia Basin – Puget Sound Research Conference, March 26–29, 2007, Vancouver, BC, Canada. Retrieved from: https://apps.dtic.mil/sti/pdfs/ADA519149.pdf. April 19, 2021.

McIntyre J, Baldwin DH, Meador JP, Scholz NL (2008) Chemosensory deprivation in juvenile coho salmon exposed to dissolved copper under varying water chemistry conditions. *Environmental Science & Technology* 42: 1352–1358.

Michalsen DR, Brown SH (2019) *Feasibility of Nearshore Placement Near the Swinomish Navigation Channel: Puget Sound, Washington.* ERDC/TN RSM-19-6. U.S. Army Engineer Research and Development Center, Vicksburg, MS. Retrieved from: https://apps.dtic.mil/sti/pdfs/AD1079678.pdf. June 4, 2021.

Moore SJ, Mantua NJ, Newton JA, Kawase M, Warner MJ, Kellogg JP (2008) A descriptive analysis of temporal and spatial patterns of variability in Puget Sound oceanographic properties. *Estuarine, Coastal and Shelf Science* 80: 545–554.

Moore JW, Scheuerell MD, Frodge J (2003) Lake eutrophication at the urban fringe, Seattle region, USA. *Ambio* 32: 13–18.

Morley SA, Toft JD, Hanson KM (2012) Ecological effects of shoreline armoring on intertidal habitats of a Puget Sound urban estuary. *Estuaries and Coasts* 35: 774–784.

Neff JM, Stout SA, Gunster DG (2005) Ecological risk assessment of polycyclic aromatic hydrocarbons in sediments: Identifying sources and ecological hazard. *Integrated Environmental Assessment and Management* 1: 22–33.

Neiman PJ, Schick LJ, Ralph FM, Hughes M, Wick GA (2011) Flooding in Western Washington: The connection to atmospheric rivers. *Journal of Hydrometeorology* 12: 1337–1358.

Nelson AR, Kelsey HM, Witter RC (2006) Great earthquakes of variable magnitude at the Cascadia subduction zone. *Quaternary Research* 65: 354–365.

O'Neill SM, West, JE (2009) Marine distribution, life history traits, and the accumulation of polychlorinated biphenyls in Chinook salmon from Puget Sound, Washington. *Transactions of the American Fisheries Society* 138: 616–632.

Offenhuber D, D Lee, MI Wolf, S Phithakkitnukoon, A Biderman, C Ratti (2012) Putting matter in place: Measuring tradeoffs in waste disposal and recycling. *Journal of the American Planning Association* 78(2): 173–196.

Palermo MR, Clausner JE, Channell MG, Averett DE (2000) *Multiuser Disposal Sites (MUDS) for Contaminated Sediments from Puget Sound—Subaqueous Capping and Confined Disposal Alternatives.* ERDC TR-00-3, U.S. Army Engineer Research and Development Center, Vicksburg, MS. Retrieved from: https://apps.dtic.mil/sti/pdfs/ADA380336.pdf. April 23, 2021.

Papenfus M (2019) Do housing prices reflect water quality impairments? Evidence from the Puget Sound. *Water Resources and Economics* 27: 100133.

Paulson AJ, Keys ME, Scholting KL (2010) *Mercury in Sediment, Water, and Biota of Sinclair Inlet, Puget Sound, Washington, 1989–2007.* U.S. Geological Survey, Washington. p. 220.

Penttila D (2007) *Marine Forage Fishes in Puget Sound.* Puget Sound Nearshore Partnership Report No. 2007-03. U.W. Army Corps of Engineers, Seattle, Washington.

Ponti DJ, Tinsley JC III, Treiman JA, Seligson H (2008) *Ground Deformation.* U.S. Geological Survey, Washington.

Prudhomme C, Young A, Watts G, Haxton T, Crooks S, Williamson J, Davies H, Dadson S, Allen S (2012) The drying up of Britain? A national estimate of changes in seasonal river flows from 11 Regional Climate Model simulations. *Hydrological Processes* 26(7): 1115–1118. DOI: 10.1002/hyp.8434.

Ralph FM, Dettinger MD (2011) Storms, floods, and the science of atmospheric rivers. *Eos* 92: 265–272.

Ross PS, Jeffries SJ, Yunker MB, Addison RF, Ikonomou MG, Calambokidis JC (2004) Harbor seals (*Phoca vitulina*) in British Columbia, Canada, and Washington State, USA, reveal a combination of local and global polychlorinated biphenyl, dioxin, and furan signals. *Environmental Toxicology and Chemistry* 23: 157–165.

Ruckelshaus MH, McClure MM, the Sound Science Collaborative Team (2007) *Sound Science: Synthesizing Ecological and Socioeconomic Information About the Puget Sound Ecosystem.* U.S. Dept. of Commerce, National Oceanic & Atmospheric Administration (NMFS), Northwest Fisheries Science Center, Seattle, Washington. p. 93.

Scholz NL, Myers MS, McCarthy SG, Labenia JS, McIntyre JK, Ylitalo GM, et al. (2011) Recurrent die-offs of adult coho salmon returning to spawn in Puget Sound lowland urban streams. *PLoS One* 6(12): e28013.

Schoof RA, DeNike J (2017) Microplastics in the context of regulation of commercial shellfish aquaculture operations. *Integrated Environmental Assessment and Management* 13: 522–527.

Schulz WH (2007) Landslide susceptibility revealed by LIDAR imagery and historical records, Seattle, Washington. *Engineering Geology* 89: 67–87.

Seattle (2018) *2018 Waste Prevention and Recycling Report.* Seattle.gov. Retrieved from: https://www.seattle.gov/Documents/Departments/SPU/Documents/Recycling_Rate_Report_2018.pdf. April 20, 2021.

Seattle (nd) *Seattle Office of Emergency Management.* City of Seattle. Retrieved from: https://www.seattle.gov/Documents/Departments/Emergency/PlansOEM/SHIVA/2014-04-23_Earthquakes(0).pdf. April 15, 2021.

SEPA (2021) *Blair Waterway Berth Maintenance Dredge.* Port of Tacoma. Retrieved from: https://www.portoftacoma.com/sepa-action-blair-waterway-berth-maintenance-dredge. April 18, 2021.

Shaffer JA, Parks DS (1994) Seasonal variations in and observations of landslide impacts on the algal composition of a Puget Sound nearshore kept forest. *Botanica Marina* 37: 315–323.

Sheets B (2012) *Kimberly-Clark Mill Leaves a Toxic Mess Behind.* HeraldNet. Retrieved from: https://www.heraldnet.com/news/kimberly-clark-mill-leaves-a-toxic-mess-behind/. April 24, 2021.

Shipman H (2010) The geomorphic setting of Puget Sound: Implications for shoreline erosion and the impacts of erosion control structures. In: Shipman H, Dethier MN, Gelfenbaum G, Fresh KL, Dinicola RS (editors). *Puget Sound Shorelines and the Impacts of Armoring – Proceedings of a State of the Science Workshop*, pp. 19–34. U.S. Geological Survey, Washington.

Snow JA, Dennison JB, Jaffe DA, Price HU, Vaughan JK, Lamb B (2003) Aircraft and surface observations of air quality in Puget Sound and a comparison to a regional model. *Atmospheric Environment* 37: 4019–4032.

Sobocinski KL, Cordell JR, Simenstad CA (2010) Effects of shoreline modifications on supratidal macroinvertebrate fauna on Puget Sound, Washington beaches. *Estuaries and Coasts* 33: 699–711.

Sotirov M, Memmler M (2012) The Advocacy Coalition Framework in natural resource policy studies—Recent experiences and further prospects. *Forest Policy and Economics* 16: 51–64.

Soundkeeper (nd) *Pollution in Our Waterways*. Puget Soundkeeper. Retrieved from: https://pugetsoundkeeper.org/pollution/. April 24, 2021.

Spromberg JA, Baldwin DH, Damm SE, McIntyre JK, Huff M, Sloan CA, Anulacion BF, Davis JW, Scholz NL (2016) Coho salmon spawner mortality in western US urban watersheds: bioinfiltration prevents lethal storm water impacts. *Journal of Applied Ecology* 53: 398–407.

Stewart J, Gast R, Fujioka R, Solo-Gabriele H, Meschke JS, Amaral-Zettler L, del Castillo E, Polz MF, Collier TK, Strom MS, Sinigalliano CD, Moeller PDR, Holland AF (2008) The coastal environment and human health: microbial indicators, pathogens, sentinels and reservoirs. *Environmental Health* 7(Suppl 2): S3. DOI: 10.1186/1476-069X-7-S2-S3.

Susewind K (2011) *Combined Sewer Overflow Programs: Protecting Our Waters from Stormwater, Raw Sewage and Industrial Pollution*. Washington State Department of Ecology Blog. Retrieved from http://ecologywa.blogspot.com/2011/07/combined-seweroverflow-programs.html. April 21, 2021.

Syphard AD, Bar Massada A, Butsic V, Keeley JE (2013) Land use planning and wildfire: development policies influence future probability of housing loss. *PLoS One* 8: e71708.

Takesue RK, Campbell PL, Conn KE (2019) *Polycyclic Aromatic Hydrocarbons, Polychlorinated Biphenyls, and Metals in Ambient Sediment at Mussel Biomonitoring Sites, Puget Sound*. U.S. Geological Survey, Washington. p. 15. DOI: 10.3133/ofr20191087.

Vaccaro JJ, Hansen Jr AJ, Jones MA (1998) *Hydrogeologic Framework of the Puget Sound Aquifer System, Washington and British Columbia*. U.S. Geological Survey, Washington. Retrieved from: https://pubs.usgs.gov/pp/1424d/report.pdf. April 23, 2021.

Van Wagoner TM, Crosson RS, Creager KC, Medema G, Preston L, Symons NP, Brocher TM (2002) Crustal structure and relocated earthquakes in the Puget Lowland, Washington, from high-resolution seismic tomography. *Journal of Geophysical Research* 107(B12): 2381.

Voosen P (2020) A muddy legacy. *Science* 369: 898–901.

Warrick JA, George DA, Gelfenbaum G, Ruggiero P, Kaminsky GM, Beirne M (2009) Beach morphology and change along the mixed grain-size delta of the dammed Elwha River, Washington. *Geomorphology* 111: 136–148.

West JE, O'Neill SM, Ylitalob GM, Incardona JP, Dotya DC, Dutch ME (2014) An evaluation of background levels and sources of polycyclic aromatic hydrocarbons in naturally spawned embryos of Pacific herring (*Clupea pallasii*) from Puget Sound, Washington, USA. *Science of the Total Environment* 499: 114–124.

Wilber D, Clarke D (2007) Defining and assessing benthic recovery following dredging and dredged material disposal. Proceedings XXVII World Dredging Congress, Orlando, FL, pp. 603–618.

Wills CJ, Perez FG, Gutierrez CI (2011) *Susceptibility to Deep-Seated Landslides in California*. California Geologic Survey. Retrieved from: https://www.conservation.ca.gov/cgs/Documents/Publications/Map-Sheets/MS_058.pdf. January 4, 2021.

Wilson RC, Keefer DK (1985) Predicting areal limits of earthquake-induced landsliding. In: JI Ziony (editor). *Evaluating Earthquake Hazards in the Los Angeles Region-An Earth-Science Perspective*, pp. 317–345. U.S. Geological Survey, Washington.

Worm B, Heike K, Lotze HK, Jubinville I, Wilcox C, Jambeck J (2017) Plastic as a persistent marine pollutant. *Annual Review of Environment and Resources* 42: 1–26.

Young AP, Olsen MJ, Driscoll N, Flick RE, Gutierrez R, Guza RT, Johnstone E, Kuester F (2010) Comparison of airborne and terrestrial Lidar estimates of seacliff erosion in Southern California. *Photogrammetric Engineering & Remote Sensing* 76: 421–427.

Zauscher MD, Wang Y, Moore MJK, Gaston CJ, Prather KA (2013) Air quality impact and physicochemical aging of biomass burning aerosols during the 2007 San Diego wildfires. *Environmental Science & Technology* 47: 7633–7643.

Zeeman C, Taylor SK, Gibson J, Little A, Gorbics C (2008) *Characterizing Exposure and Potential Impacts of Contaminants on Seabirds Nesting at South San Diego Bay Unit of the San Diego National Wildlife Refuge (Salt Works, San Diego Bay)*. Final Report. United States Fish and Wildlife Service, Carlsbad.

8 Biology of Puget Sound

In this chapter, we describe the contemporary inhabitants of the Puget Sound region, in particular the trees, vegetation, marine and estuarine plants and algae, as well as the major groups of animals, such as the keystone predators (e.g., mountain lions and coyotes on land, sea lions and whales in the sea) and their prey (deer, wild pigs, smaller predators, rodents, and reptiles). Puget Sound is also a major region for overwintering and migrating birds and butterflies [hummingbirds, e.g., the black-chinned hummingbird (*Archilochus alexandri*), Calliope (*Selasphorus calliope*), and Rufous (*Selasphorus rufus*)] and the main inflows of the Snohomish, Skagit, and Puyallup Rivers serve as sources of fresh water for the region's biota. Almost all the animal groups are discussed, including vertebrates and invertebrates, both terrestrial and aquatic, as well as the region's botanical denizens. We also discuss the contemporary problems with the numerous invasive species that have been introduced, both deliberately and accidentally, by man over the past 200 years or so.

INTRODUCTION

The Puget Sound region has predominantly a warm Mediterranean climate (Csb) and a warm humid climate (Cfb) in the mountainous regions to the west and east, when using the Köppen climate classification system (Kottek et al., 2006). The climate is characterized by cool mild winters and warm dry summers, with most of the precipitation (more than 900 mm) occurring during the winter months (Federal Records, 2021; NOAA, 2011). The flora and fauna are typical of such a climate zone and many of the species endemic to the region are also to be found in similar climate zones in Eurasia (Jiang et al., 2019).

Regarding a description of the flora and fauna of the Puget Sound region, it is not our intention to provide an encyclopedic list of all the resident species. Nor should this section be relied upon as a definitive accounting of the flora and fauna of the region. Rather, we will give a few chosen examples of how the environment has affected (and sometimes changed) the evolution of various organisms. As in Chapter 5 we will use the Linnaean binomial nomenclature where possible for additional study by the reader.

ANIMALS

VERTEBRATES

This section includes descriptions of all the extant animals that are classified as chordates, implying they have had at least a dorsal notochord and/or a nerve cord during ontogeny (embryonic growth and development) (Rychel et al., 2006).

DOI: 10.1201/9780429487439-8

As described in Chapter 5, the fauna of the Upper Pleistocene Epoch (129 to 11.65 kya) (kya: thousand years ago) were predominantly mammoths, mastodons, shrub ox, woolly rhinoceros, giant sloths, horses, camels, llamas, the monstrous short-faced bear, a number of large saber-toothed cats, American lion, American cheetah, and the dire wolf, in addition to the ancestors of those that survived into our present time, the Holocene Epoch (11.65 kya to the present).

We will introduce the description of the fauna of the Puget Sound region by way of class, size, and behavioral groupings. We will only include those animals that have in the past or currently occupy ecological niches in the region; we may also mention if a species that is considered endemic (a species confined to a specific ecological or geographical area) to Puget Sound is also found elsewhere, although this is very rare. We will describe in classic trophic terms, from the top predators through the herbivores and those smaller animals that depend upon both groups.

MAMMALS

Class Mammalia are descendants of the group of Mesozoic mammal-like reptiles, the therapsids, and which in fact significantly pre-date the better-known dinosaurs of the time (Kemp, 2005); they most likely evolved as incompletely endothermic reptiles from the pelycosaurs of the Permian Period (299 to 252 mya) (mya: million years ago) in response to a warming and dry climate (Kemp, 1987, 2005). As many readers will likely know, during the Cretaceous Period (145 to 66 mya), the ancestors of the extant mammals were usually quite small creatures, often living in burrows and being more active at night (Cifelli, 2000; Woodburne, 2004). Mammals are members of the clade Amniota, which are characterized by having a membrane, the amnion, surrounding the developing embryo. The ancestors of some present-day mammals in northwest Washington were described in more detail in Chapter 5.

PREDATORS

We first come to the Carnivora, the carnivores, which in turn comprise the following Families: Ursidae (bears), Canidae (dogs and their relatives), Mustelidae (weasels, martens, otters, badgers, skunks), Procyonidae (raccoons), Otariidae (eared seals), Phocidae (earless or true seals), and Felidae (all cats) (Wilson and Mittermeier, 2009; Flynn et al., 2005). The large or largest predators in an ecosystem are referred to by biologists and ecologists as a keystone predator. In this context, the term "keystone" refers to the influential niche position of the predator within the ecosystem and which contributes to how the ecosystem is maintained in balance (Beschta and Ripple, 2009; Wallach et al., 2010; Ripple et al., 2014). This is referred to as a top-down trophic cascade (Leopold, 1949; Hairston et al., 1960; Oksanen et al., 1981). Briefly, a keystone predator maintains an ecosystem in balance by predation upon herbivores that themselves feed upon vegetation critical to maintaining the structure of the ecosystem; for example, by securing the banks of a stream or river so that flooding and sedimentary buildup is controlled. If the keystone predator is removed, such as by human control measures or by disease, the prey herbivores multiply uncontrollably, devour more vegetation, and with less vegetation to secure the banks of the stream or river, the fluvial flow is compromised, leading to flow rate changes, build-up of silt, and eventual change of the topography and geomorphology of the affected area (Beschta and Ripple, 2009). Most times, the keystone predator is also one of the apex predators in the ecosystem.

Three large keystone predators survived the great megafauna extinction as discussed in Chapter 5: the grizzly bear (*Ursus arctos horribilis*), the mountain lion (*Puma concolor*), and the gray wolf (*Canis lupus*). The grizzly bear was listed as threatened under the US federal Endangered Species Act on July 28, 1975. Since that listing, the US Fish and Wildlife Service initiated a recovery effort to be managed in six recovery zones in the western US (in conjunction with its Canadian partner, the Canadian Wildlife Service): one of these is the North Cascades Ecosystem (NCE) which straddles the US-Canada border bounded in the north by the Fraser River, the east by the Okanogan Highlands and Columbia Plateau, to the south by Snoqualmie Pass, and the Puget lowlands to the west (USFWS, 2013). There have only been about four sightings of grizzlies in the NCE over the past 15 years and population may only amount to two (USFWS, 2013). In Canada, the northwestern population (including British Columbia) was designated Special Concern in April 1991 and populations in British Columbia may total around 14,000 with the majority inhabiting the Columbia and Rocky Mountains in the southeast of the province (COSEWIC, 2002). Recent estimates for the North Cascades grizzlie population number only in the 20s having no biogeographic access to other populations in British Columbia (COSEWIC, 2002). The US Fish and Wildlife Service has cited the NCE as one of the most intact wildlands in the United States and proposed that the most plausible carrying capacity for the US portion of the NCE is approximately 280 bears (Lyons et al., 2016; USFWS, 2013).

Although the large keystone predators (grizzly bear, *Ursus arctos horribilis*) and gray wolves, (*Canis lupus*) had been hunted almost to extinction by the mid-1960s (Sullivan, 1983) there still remain a number of keystone predators in the region. At one time, keystone predators were considered to be detrimental to livestock and danger to people, and so during most of the 19th and 20th centuries they were allowed to be hunted at will; in addition, there was a huge demand for bear hides for the fur trade throughout the 19th century, according to records kept by the Hudson's Bay Company with nearly 4000 accounted for (Sullivan, 1983). During the late 1980s and onwards, ecologists and other biologists determined that removing these keystone predators from the ecosystem led to significant damage to the environment (Edvenson, 1994; Beschta and Ripple, 2009).

Scientists now consider that the grizzly bear is really just a larger brown bear (also *Ursus arctos*) that migrated from Eurasia after the last Ice Age about 14 kya (thousand years ago) as they have no significant mitochondrial DNA differences (Cronin et al., 1991).

The remaining and extant large predators in the Pacific Northwest include the American black bear (*Ursus americanus altifrontalis*) inhabiting the north coast and the Cascades, mountain lion (*Puma concolor*), also known as the puma, cougar, or catamount (probably an Anglicization of the Spanish for mountain lion, "gato monte"), the gray wolf (*Canis lupus*), most likely migrating south from Canada, and the coyote (*Canis latrans*).

In the Puget Sound region, cougars and coyotes have become more tolerant to humans and encounters with either species within municipal boundaries are often reported in the local news (Wagner, 2021; Lim, 2021). Residents at the periphery of the urban and suburban communities may often hear the squealing and yipping of coyote pups greeting their parents return after a successful night's hunt during the mid-summer.

Another more-rarely spotted predator limited to alpine meadows of the Cascades is the Cascade red fox (*Vulpes cascadensis*) and these too have become habituated to urban and suburban life. Interestingly, these red foxes are part of the fox clades that had inhabited the southern (or montane) refugium (subalpine parklands and alpine meadows of the Rocky Mountains, the Cascade Range, and Sierra Nevada) during the last ice age, and are only distantly related to the North American eastern red fox (*Vulpes vulpes*) which is more often seen at low elevations. These are descended from foxes that escaped from fur farms during the early to mid-1900s. It is considered to be more of a problem than the Cascade red fox, as it indiscriminately hunts small birds, small rodents, and reptiles whereas the Cascade red fox is more in harmony with the ecosystem it has evolved with, hunting rodent pests, lagomorphs, and insects; it is also frequently herbivorous (Fedriani et al., 2000; Aubry et al., 2009; Sacks et al., 2010; Sacks et al., 2011).

The mountain lion (*Puma concolor*), also known as the puma or cougar, is to be found throughout the Cascades and there are also small populations found in the Olympic National Park, as well as the Puget Lowlands (NPS, ND; Brehme et al., 2014). Their natural prey are deer, elk, moose, mountain goats, wild sheep, coyotes, galliform birds, such as turkey (*Meleagris gallopavo*), as well as rodents and raccoons (WDFW, NDb).

The bobcat (*Lynx rufus*) is up to two times larger than the domestic cat, standing 38 cm (15"), and weighs about 9 kg (20 lb; females) and between 7 and 13.5 kg (16–30 lb; males) (CDFW, ND). They are present throughout the Puget Sound region but as they are generally nocturnal, are observed infrequently. During the day, to avoid larger predators (and humans) they rest in what is termed a "scrape", which is a shallow pit scraped out with their hind limbs. The bobcat is most likely descended from *L. issiodorensis*, whose fossils have been found in western Europe, migrated across the Bering Land Bridge (Beringia) about 2.6 mya (Johnson and O'Brien, 1997; Pecon-Slattery and O'Brien, 1998), but then became isolated during the subsequent Ice Ages. Its close relative, the Canada lynx (*L. canadensis*) is to be found farther east in the Northern Cascades and are most likely is a descendent of a second invasion of the Eurasian lynx (*L. lynx*) within the last 200 ky; fossils of *L. issiodorensis* have been found (Tumlison, 1987; Anderson and Lovallo, 2003). Some interesting studies have been performed on the population fluctuations of the Canada lynx and its main prey, the snowshoe hare (*Lepus americanus*) (Hunter, 2015). A review of the two populations' furs traded by the Hudson's Bay Company from 1845 to 1936 showed a cyclical and parallel boom and bust of between 3- and 17-fold changes in the predator and prey populations (Brand and Keith, 1979; Ward and Krebs, 1985; Poole, 2003): there appears to be an 8- to 11-year periodic cycle, with the lynx population cycling about 1–2 years behind the snowshoe hare population (Weinstein, 1977). This ecological phenomenon has been cited as textbook examples of the Lotka–Volterra predator–prey equation, demonstrating the interplay of three major factors—food, predation, and social interaction (Krebs et al., 2001).

The Musteloidea are a superfamily of the Order Carnivora and to which belong many of the small predators common in the Puget Sound region. They include the Procyonidae (raccoons and their allies, named for their original classification as "pre-dogs"), the Mustelidae (otters, badgers, martens, weasels and their allies, named for the musk (scent/must) gland used to mark territory), the Mephetidae (skunks), and the Ailuridae (the red panda).

The common raccoon *(Procyon lotor)* is the largest and the most prevalent member of the procyonid family in North America, but unlike most other carnivores, it is an opportunistic omnivore (Wozencraft, 2005). They are defined as a mesopredator, which are characterized by the rapid population growth of intermediate-sized predators in the absence of larger top predators (Martin, 2011).

The name "raccoon" is derived from the Algonquian word *aroughcoune*, meaning "he who scratches with his hands." The Algonquian nation (who call themselves *Omàmiwininiwak* and *Anicinàpe*) is in northeastern North America (present-day Québec and eastern Ontario) and who had traded raccoon furs with the French during the 16th to 18th centuries (Litalien, 2004; Poulter, 2010). Raccoons are closely related to the ringtail (*Bassariscus* sp.), found in the dryer parts of Oregon, Eastern California, and extending to the southwestern United States, the olingos (*Bassaricyon* sp.), the coati (*Nasua* and *Nasuella* spp.), and more distantly related to the kinkajou (*Potos flavus*), a species only found in Central and South America (Wozencraft, 2005). The common raccoon is largely nocturnal, digging up worms and small invertebrates, and small family groups are often seen wandering the suburbs of Seattle and other conurbations. Whilst they may hibernate during the winter in other parts of North America, the Pacific Northwest climate is sufficiently mild that raccoons active throughout the year. Although other procyonids are generally form social groups, raccoons differ in that, other than sow-kit groups, they tend to remain solitary and may engage in fights during encounters with other raccoons (Barrat, 2013).

Members of the family Mustelidae including, otters (both sea and river subspecies), martens, and weasels, are common throughout the Puget Sound region (Brehme et al., 2014; Lafferty and Tinker, 2014); these are also termed mesopredators and we will now describe them in more detail.

The sea otter (*Enhydra lutris*) is native to the west coast of North America and is distinct from the Eurasian otter (*Lutra lutra*). It is the heaviest of the mustelids and unlike the others, has no scent gland (Kenyon, 1969). Sea otters make their home amongst the giant kelp forests growing in the shallow waters of the coast (see above) and its shelter provides much of the food for them, mainly echinoderms, such as sea urchins, crustaceans, mollusks, and fish, and it is a keystone species. A close relative, the North American river otter (*Lontra canadensis*), inhabits fresh water, and is often seen in Puget Sound and the neighboring waterways. Its prey differs from that of the sea otter in that it principally consumes fish, amphibians, mollusks, and crustaceans (Grenfell, 1974; Melquist and Dronkert, 1987; Larsen, 1984).

The Pacific marten (*Martes caurina*) was believed to be almost extinct on the Olympic Peninsula by the early part of the 21st century; however, motion-triggered cameras have recently shown them to be present as a very small population. Also spotted on camera was their larger relative, the fisher (*Pekania pennanti*) which were reintroduced to the peninsula in 2008. Martens are more common throughout the Cascade ranges (WPZ, ND).

The wolverine (*Gulo gulo luscus*) is the largest of the North American land-dwelling mustelids; it is primarily a scavenger of carrion and its diet includes almost every mammal within its range, including moose and elk (Scrafford and Boyce, 2018). Genetic evidence suggests that it likely originated from a single Eurasian population that was isolated in and then migrated from Beringia following the last Ice Age, similar to the incoming migrations made by humans (Tomasik and Cook, 2005).

American mink (*Neovision vision*) has been prized for centuries for its luxurious fur with fur farms established in both the Americas and Europe. They prey on fish, small mammals, and birds; they are hunted by gray wolves, coyotes, foxes, bobcats, and great horned owls (ADFG, ND). They inhabit throughout the state but prefer to live close to waterways (UWCAS, ND).

Another common mustelids resident in the Puget Sound region is the long-tailed weasel (*Mustela frenata*) and which is generally found in the more rural parts of the Pacific Northwest.

The striped skunk (*Mephitis mephitis*) was once included within the mustelids, but is now classified in its own family, Mephetidae, and like the raccoon, is an omnivore (Dragoo and Honeycutt, 1997). Its territory is usually confined to brushy and open country. It is primarily nocturnal although it can be easily detected during the daylight hours by its pungent smell, originating in its anal scent glands, situated ventral to its tail. Another skunk, the Western spotted skunk (*Spilogale gracilis*) shares a similar range in the region, however, prefers to inhabit mixed woodlands, farmlands, and open areas and so does not always compete directly with its relative (UWCAS, ND).

A visitor to the Seattle waterfront, the marinas, or the waterways of the eastern bays is unlikely to miss the large marine predator, the California sea lion (*Zalophus californianus*). It is in the family Otariidae, the eared seals, having external ear flaps, and which differ from true seals, which lack external ears. They are also distinguished from the true seal in that their front flippers are long and strong enough to hold themselves upright and they have rear flippers which are used for propulsion on land, facing forwards (Jeffries et al., 2000).

The Steller sea lion (*Eumetopias jubatus*), also in the family Otariidae, is a less-frequent visitor to populated areas on the Pacific Northwest, preferring quieter areas of Puget Sound and the Salish Sea (Jeffries et al., 2000). They are considered "Threatened" both at the State and Federal level (UWCAS, ND).

Another rare visitor to the Puget Sound region is the northern fur seal (*Callorhinus ursinus*), another in the family Otariidae, but these rarely come ashore on the mainland and are mainly found on the Washington west coast, Vancouver Island or out at sea, hunting. They are a protected species under the Marine Mammal Protection Act of 1972 (Jeffries et al., 2000).

True seals, on the other hand, are represented on the Puget Sound region coastline by the Pacific harbor seal (*Phoca vitulina richardsi*). They are more specialized for aquatic life than are sea lions, having lost the ability to ambulate using their flippers on land. They are the predominant species of seal seen in Puget Sound proper and the San Juan Islands (Jeffries et al., 2000).

Another visitor to the Puget Sound and the Salish Sea is the northern elephant seal (*Mirounga angustirostris*) and which has become a more common sight during the past 30–40 years in Washington than in the past (Abadía-Cardoso et al., 2017) when they became protected under the Marine Mammal Protection Act. Since then, their numbers have increased spectacularly and have probably reached the population size before they became targets for their oil. They have been seen with newborn pups on beaches on the northeastern coast of the Olympic Peninsula (Jeffries et al., 2000).

In the last century, the Pacific gray whale (*Eschrichtius robustus*) was hunted nearly to extinction. Between 1846 and 1874, as many as 8000 gray whales were killed by whalers. Other hunts continued until 1936 when gray whales became protected by the US (Rice, 1998). By the 1940s, there was sufficient concern amongst the scientific and many western industrial communities that a moratorium on hunting should be enacted. The International Whaling Commission (IWC) enacted such a moratorium in 1946 and 1982, although some nations and aboriginal communities were permitted to continue at much-reduced levels ("The Economist" article, 2012). Since those times, baleen whale populations have increased significantly and residents of the Pacific Northwest, and occasionaly in Puget Sound proper, are able to observe up close migrating gray whales (*Eschrichtius robustus*), blue whales (*Balaenoptera musculus*), fin whales (*Balaenoptera physalus*), and humpback whales (*Megaptera novaeangliae*); rarely-seen non-migratory minke whales (*Balaenoptera acutorostrata*) and sperm whales (*Physeter catodon*), may be observed at many times of the year (WDFW, 2021a). More rarely, humpback whales (*Megaptera novaeangliae*) are seen. The Black right whale (*Eubalaena glacialis*) is presumed at or near extirpation having last been seen in 1992 (WDFW, 2021a).

Perhaps emblematic of Puget Sound to many visitors is the orca (*Orcinus orca*), particularly the southern resident population which has graced the Puget Sound region, San Juan Islands, and eastern Strait of Juan de Fuca for many decades (Balcomb and Bigg, 1986). Unfortunately, this population has recently been falling in number and it is believed that this may be due to recent drastic reduction in the chinook salmon returning to their spawning headstreams as well as increased marine water temperatures (Mapes, 2019).

In addition to the magnificent immense baleen whales, smaller toothed whales, such as dolphins (Delphinidae) and porpoises (Phocoenidae) are also frequently observed coursing back and forth along the coast, preying upon schools of fish, usually sardines, anchovy, or squid. The Pacific white-sided dolphin (*Lagenorynchus abliquidens*), Dall's porpoise (*Phocoenoides dalli*), and the harbor porpoise (*Phocoena phocoena*) are typical examples seen in the Salish Sea and Puget Sound (WDFW, 2021).

PREY

ARTIODACTYLA: EVEN-TOED UNGULATES

The vast majority of large prey mammals familiar to residents of the Pacific Northwest are deer, predominantly the Columbian black-tailed deer (*Odocoileus hemionus columbianus*) and which normally inhabit a 100-km-wide band of forested mountains of the Olympic Peninsula extending inland from the Pacific Ocean as well as the western Cascades. They are often mistakenly called "mule deer" but are a separate subspecies related to the California mule deer (*Odocoileus hemionus californicus*) that is found on the eastern flanks of the Cascades and in the hills and mountains in eastern Washington State, so called for their large ears.

Rooseveld elk (*Cervus canadensis roosevelti*) are a subspecies of the wapiti, which found over most of North America, from New Mexico in the south to British Columbia in the north, with smaller herds in Texas, North Dakota, and southern Canada. The name "wapiti" means "white rump" in the Shawnee and Cree languages, which is often all you would see of the animals in the wild (Canadian Encyclopedia). Roosevelt elk are found in the forests of the Olympic Peninsula and the Cascades (Figure 8.1). Rocky Mountain elk (*Cervus canadensis nelsoni*) are found in the mountain ranges and shrubsteppe of eastern Washington; many of them are descended from elk transplanted from Yellowstone National Park in the early

FIGURE 8.1   Roosevelt elk battling (Photograph courtesy of J. Preston and the National Park Service).

1900s. They are the typical prey species for the black bear, gray wolf, and the cougar (WDFW, 2021a).

The moose (*Alces alces*) is represented in the northern Cascades by a sub-species, the Shiras moose, which is physically smaller than the more northerly and eastern moose (*Alces americanus*) in Canada and northeastern North America; the majority of the 5000 or so moose in Washington are in the Selkirk Mountains in the northeastern part of the state and the high desert eastern Columbia Basin (WDFW, 2021a). Interestingly, moose (including those in Eurasia), reindeer/caribou, and roe deer (a small deer found in Europe, western and southern Asia) belong to the deer subfamily Capreolinae (telemetacarpal deer), which are distinguished from all the other North American deer, the Cervinae, by having a distal lateral metacarpal [a bony remnant of one of the lost forelimb metacarpal adjacent to the severely reduced phalange (toe-bone)] whereas the Cervinae (plesiometacarpal deer) have a proximal lateral metacarpal (bony remnant of the lost forelimb adjacent to the carpal [wrist bone]) (Azanza et al., 2013).

Other artiodactyls common to the greater Puget Sound region are mountain goats (*Oreamnos americanus*), native to the North Cascades and introduced to the Olympic Peninsula ranges, and Bighorn sheep (*Ovis canadensis*), common predominantly in the North Cascades and both preferring rocky areas with scattered trees (WDFW, 2021a)

European wild boar (*Sus scrofa scrofa*), predominantly descendants of a Eurasian and island Southeast Asian wild pigs are also artiodactyls, may have been introduced into "game farms" on the Olympic Peninsula for hunting during the 1930s; its domesticated descendant, common swine or pigs were brought over by Spanish settlers during the 1700s and many have since become feral and the present-day population is a wild boar/swine hybrid (Larson et al., 2005; Chen et al., 2007; Rouhe and Sytsma, 2007; Woodward and Quinn, 2011; Mayer and Brisbin, 2008; Mayer, 2017; USDA, 2020). Their populations are generally lo-cated south of the Olympic Peninsula in Lewis County and the Quinault Indian Reservation and are not common in the Puget Sound region.

Perissodactyla: Odd-Toed Ungulates

North America was once home to perissodactyls, which include horses, rhino-ceroses, and tapirs, but the only remaining species native to the Americas are the Central and South American tapirs (*Tapirus* spp.). As discussed in Chapter 5, the horse (*Equus ferus/caballus sp.*) evolved in North America during the late Pliocene Epoch (3.6 to 2.58 mya) but became extinct by about 10,000 ya, after the end of the last Ice Age. Today's wild horses of the North American plains and desert mountain ranges are descendants of horses brought during the various conquests by Europeans, a large proportion related to Marismeño horses from the salt marshes of the Guadalquivir River on the Iberian Peninsula (Luís et al., 2006).

Mid-sized and small prey mammals, usually rodents, are commonly seen in all parts of the Puget Sound region. Some typical examples are: the Cascade golden-mantled ground squirrel (*Callospermophilus saturatus*) probably isolated from its southern and eastern relative, the golden-mantled ground squirrel (*Callospermophilus lateralis*), by the Columbia River (Eiler and Banack, 2004); the Douglas squirrel (*Tamiasciurus douglasii*) and the Western gray squirrel (*Sciurus griseus nigripes*), the Northern flying squirrel (*Glaucomys sabrinus*), all inhabitants of woodlands and forests of western Washington; the fox squirrel (*Sciurus niger*) and Eastern gray squirrel (*Sciurus carolinensis*), both of which appear to have been introduced to the Pacific Northwest from the east coast around 1915 (Yocom, 1950; WDFW, NDa; King, 2004; Ortiz and Muchlinski, 2014); the Eastern gray squirrel is now is the predominant species in the Puget Sound urban regions over the Western gray squirrel. Common woodland and forest floor rodents in the Puget Sound region are Townsend's chipmunk (*Neotamias townsendii*) and the yellow-pine chipmunk (*Neotamias amoenus*), the latter restricted to the Cascades east of Puget Sound. The largest ground squirrel in North America is the hoary marmot (*Marmota caligata*) and its habitat is the Cascade ranges throughout the Pacific Northwest; it is nicknamed "the whistler" for its high-pitched warning to alert others of possible danger. Other marmot denizens of the higher Cascades, the Olympic Peninsula, and Vancouver Island are the yellow-bellied marmot (*M. flaviventris*), the Olympic marmot (*M. olympus*), and the Vancouver Island marmot (*M. vancouverensis*), respectively (Kruckenhauser et al., 1999; Steppan et al., 1999).

Other rodents include native bushy-tailed woodrats (*Neotoma cinerea*), the original "pack-rat" (BMNHC, 2006; Randall, 2007); the brown rat (*Rattus norvegicus*, introduced 1750–1755; Norwak, 1999); the more common black rat (*Rattus rattus*), observed often in yards and may invade attic space (so-called "roof rats"), which was probably introduced during the early- to mid-1800s and whose mtDNA is more closely related to rats from South and South-east Asia (Lantz, 1909; Conroy et al., 2012); the muskrat (*Ondatra zibethicus*) inhabiting Puget Sound wetlands and upland waterways; the coypu (*Mycocastor coypus*; also known as the nutria) is the third-largest rodent worldwide [behind the beaver and the capybara (*Hydrochoerus hydrochaeris*)] and is an invasive species in the Puget Sound region, originally introduced from South America; Townsend's vole (*Microtus townsendii*), which also primarily inhabits waterways throughout the Pacific Northwest; the deer mouse (*Peromyscus maniculatus*), the most populous mammal in the United States and the carrier of the notorious hantavirus and Lyme disease (CDC, 2012; CCDR, 2008); the introduced house mouse (*Mus musculus domesticus*); rabbits and their kin, including the American pika (*Ochotona [Pika] princeps*), typically found in the Cascade and Olympic mountain ranges in boulder fields or above the tree line; the snowshoe hare (*Lepus americanus*) and the eastern cottontail (*Sylvilagus floridanus*) (introduced, found mainly on Vancouver Island and eastern Cascades); shrews, including the Pacific water shrew (*Sorex bendirii bendirii [Cascades] and S. b. albiventer [Olympic peninsula]*) and Trowbridge's shrew (*Sorex trowbridgii*); moles, including the American shrew mole (*Neurotrichus gibbsii*), and the coast or Pacific mole (*Scapanus orarius*); and bats, including the little brown myotis (*Myotis lucifugus alascensis*) and the hoary bat (*Lasiurus cinereus*).

Another important large rodent that claims the northwestern United States as home is the mountain beaver, (*Aplodontia rufa rufa* in the Olympic Peninsula and *A. r. rainieri* in the Cascades) and which is distinct from the North American beaver (*Castor canadensis*); it is distantly related to squirrels (Macdonald, 2001). The North American beaver is found throughout the same geographic ranges but as they occupy different ecological niches, this limits competition (Macdonald, 2001). As with many of the small- to mid-sized mammals, their territories have been subject to continuous encroachment by people, farming, and urbanization, and thus their population numbers and range have significantly decreased over the past 250 years or so.

One bane of Puget Sound region residents is the Mazama pocket gopher (*Thomomys mazama*); in their natural habitat, the Olympic Peninsula and south of Tacoma, they create a network of tunnel systems that provide protection and a means of collecting food. However, when they invade peoples' private gardens, they can cause havoc and destroy lawns, small shrubs, and entire horticultural endeavors. A related species, the northern pocket gopher (*Thomomys talpoides*), inhabits the Cascades and farther east (Burke Museum 2021a).

MARSUPIALS

The only marsupial found in North America is the Virginia opossum (*Didelphis virginiana*). This ancient pouched mammal lives for only about three years but breeds when it is six (for a female) or eight (male) months old up to three times a year. They are generally nocturnal but are a familiar sight even in urban environments, especially when food is left outside for pets; they are omnivorous (Krause and Krause, 2006). The tail is prehensile and young opossums use this to grip onto the mother's back. As described in Chapter 5, the opossum's presence in North America resulted from the Great American Biotic Interchange (GABI) just under three million years ago.

BIRDS AND REPTILES

Class Reptilia include all the predominantly exothermic and endothermic tetrapods that usually lay eggs on land, eggs that are protected by a gas-permeable, proteinaceous and/or calciferous shell surrounding and protecting the amnion. They represent the group of tetrapods that successfully adapted and emigrated from the aquatic environment to the terrestrial environment during the Carboniferous Period, the amniotes (359 to 299 mya). Living members include lizards, snakes, crocodilians, and birds.

BIRDS

Clade Aves now is considered to be placed within the larger clade of Dinosauria, and they are generally referred to by taxonomists as avian dinosaurs (Ostrom, 1973; Padian, 1986).

Land Birds

For reasons of space, we cannot describe each bird species found in the Puget Sound region and we suggest the reader finds other more comprehensive resources, such as the many ornithological handbooks available to the general public. Here we describe some of the birds that are of particular significance to the ecology of the Puget Sound region.

The turkey vulture (*Cathartes aura*), sometimes called a turkey buzzard, is another member of the New World vultures and is the most common scavenging bird seen in the skies of the Puget Sound region; as would be expected, they are often seen soaring and circling up using updrafts high into the sky to better see over the vast plains and hills of the Pacific Northwest. They are also to be found in the middle of the highway, a small group tearing at the carcass of road-kill.

The Puget Sound region has a large number of native (and some invasive) birds of prey, in particular since the environment provides habitat for so many of their typical prey. These include the great horned owl (*Bubo virginianus*); the northern spotted owl (*Strix occidentalis caurina*) and its main competitor the invading northern barred owl (*Strix varia varia*), the western screech owl (*Megascops kennicottii*), the bald eagle (*Haliaeetus leucocephalus*) and golden eagle (*Aquila chrysaetos*), the osprey (*Pandion haliaetus carolinensis*), the prairie falcon (*Falco mexicanus*), the American peregrine falcon (*Falco peregrinus anatum*), the American kestrel or sparrow hawk (*Falco sparverius*), the red-tailed hawk (*Buteo jamaicensis*), northern goshawk (*Accipiter gentilis laingi*; a smaller subspecies found on Vancouver Island and the Haida Gwai archipelago [previously the Queen Charlotte Islands]), northern harrier or marsh hawk (*Circus cyaneus*), and the merlin (*Falco columbarius*) (Robbins et al., 1983; Sibley, 2000).

The smallest hawk in North America is the male sharp-shinned hawk (*Acciper striatus striatus*, female, 150 to 220 g); it prefers temperate mixed forest and will prey on rodents and birds, even bigger than its own size, including the ruffed grouse (*Bonasa umbellus*, 450–750 g). It lives year-round from central California to the coastal regions of Alaska (Fergeson-Leese et al., 2001).

The bald eagle was almost extinct due to heavy use of pesticide contamination of their prey, but are now more commonly seen, as they migrate to and from Canada, Montana, Wyoming, and Idaho during the spring and fall (FS, ND; Audubon, ND).

The crow family in the Puget Sound region is distinguished by the common raven (*Corvus corax*), the American crow (*Corvus brachyrhynchos*), Steller's jay (*Cyanocitta stelleri stelleri*), the Canada (or gray) jay (*Perisoreus canadensis griseus*), all of which are predominantly forest, woodland, and scrub birds (Robbins et al., 1983).

As mentioned in Chapter 4, the Puget Sound region is a critical stopover point for migratory birds along the Pacific Flyway migration route, in particular, the shallow waters of Boundary Bay (British Columbia), just north of the San Juan Islands, Puget Sound proper, and the Hood Canal estuary, which are a welcome rest point for shorebirds and waterfowl, as well as for sparrows and thrushes on the land, and which can number of up to one million birds.

The Pacific Northwest/Puget Sound regions are the wintering range of the common loon (*Gavia immer*), also known as the great northern diver in Eurasia; they eat a wide variety of animal prey, including fish, insect larvae, crustaceans, mollusks (Sandilands, 2011). Of interest to the reader, the loon's location call, a long, three-note call sounding like a low whistle, is ubiquitously used by film-makers to bring an atmosphere of mystery to a wetlands or temperate waterways scene (Wolcott, 1982).

Another bird whose winter range includes the Puget Sound region is the horned grebe (*Podiceps auritus*); it too is common in north Atlantic localities: Greenland, Iceland, Scotland, and Norway (Fjeldsa, 1973). The western grebe (*Aechmophorus occidentalis*) is the largest of the North American grebes, but it is only found in eastern Washington. The pied-billed grebe (*Podilymbus podiceps*), compared with the preceding grebes, are small, stocky, and short-necked, thereby immediately distinguishing them from other grebes (Muller and Storer, 1999).

The tundra swan (*Cygnus columbianus*), which breeds in the Arctic tundra of North America and Siberia, is a summer migrant to the Puget Sound region, where it is also known as the whistling swan (Jobling, 2010).

The American wigeon (*Mareca americana*) is a non-breeding winter visitor to Puget Sound and is most numerous of the North American Flyways on the Pacific Flyway (Floyd, 2008). American wigeon nest farther north than any other dabbling duck with the exception of the northern pintail (*Anas acuta*). The American wigeon is the fifth most commonly harvested duck in the United States, behind the mallard (*Anas platyrhynchos*), green-winged teal (*Anas carolinensis*), gadwall (*Mareca strepera strepera*, also related to the wigeon), and wood duck (*Aix sponsa*) (DU, 2010).

Some of the other native birds and waterfowl are listed here: the American goldfinch (*Carduelis tristis*), the American robin (*Turdus migratorius*), Brewer's blackbird (*Euphagus cyanocephalus*), red-winged blackbird (*Agelaius phoeniceus*), yellow-headed blackbird (*Xanthocephalus xanthocephalus*), Bullock's oriole (*Icterus bullockii*), coastal red-shafted flicker (*Colaptes auratus colluris*), spotted towhee (*Pipilo fuscus*), dark-eyed junco (*Junco hyemalis*), black-capped chickadee (*Poecile atricapillus*), boreal chickadee (*P. hudsonicus*), chestnut-backed chickadee (*P. rufescens rufescens*), American bushtit (*Psaltriparus minimus*), white-breasted nuthatch (*Sitta carolinensis tenuissima* and *S. c. aculeata*), great blue heron (*Ardea herodias fannini*), green heron (*Butorides virescens anthonyi*), scaups (*Aythya* spp.), Canada goose (*Branta canadensis*), snow goose (*Chen caerulescens*), red-breasted merganser (*Mergus serrator*, a diving duck), Bewick's wren (*Thryomanes bewickii*), and hummingbirds, including the rufous hummingbird (*Selasphorus rufus*), and the calliope hummingbird (*Selasphorus calliope*) (Robbins et al., 1983). The European or common starling (*Sturnus vulgaris*) is an invasive species, but giant flocks are to be seen in the early evenings, where they gather before nesting prior to sunset. Examples of other non-native birds are mitred conures or parakeets (*Aratinga mitrata*) found in the north Seattle area (UPS, ND).

The present Washington wild turkey population derives from wild birds re-introduced from other areas by game officials in the early 20th century. The Rio Grande wild turkey (*M. gallopavo intermedia*) was introduced into southeastern Washington during the 1960s and 1970s and Merriam's wild turkey (*M. g. merriami*) into central and northeast Washington, whereas the Eastern wild turkey (*M. g. silvestris*) is the only subspecies present in western Washington (WDFW, 2021a).

A bird frequently seen around the Pacific Northwest coastlines from spring through autumn is a subspecies of the brown pelican, the California brown pelican (*Pelecanus occidentalis californicus*); it feeds on fish by dive-bombing into the water from a height of usually about 5–10 m (WDFW, 2021a).

For further research, the reader is recommended to use some of the many well-produced handbooks available in the general press.

Seabirds and Shorebirds

One ecosystem in the Puget Sound region that we have not yet discussed is that of the peri-marine environment, the shoreline and the areas adjacent to them. It is here that numerous gulls and other sea- and shorebirds compete for food and nesting space. There are many recognized species within the gull group (Laridae) although genetic analysis often demonstrates hybridization between "species" and that most have up to 98.7% DNA in common (Paton, 2003; Pons et al., 2005; van Tuinen et al., 2004). The following lists the more common gulls seen in the Puget Sound region, not only at the seashore, but also farther inland, where they may be seen raiding trash heaps up to 200 km from the ocean (Akerman and Peterson, 2017). Included in no particular order are the California gull (*Larus californicus*), the glaucous-winged gull (*L. hyperboreus*), ring-billed gull (*L. delawarensis*), mew gull (*L. canus brachyrhynchus*), and Boneparte's gull (*Chroicocephalus philadelphia*). Some eat brine shrimp, grasshoppers, seal afterbirths, fish, squid, carrion, small mammals, and the eggs and chicks of other seabirds—even those of their own kind.

Other smaller seabirds include the common tern (*Sterna hirundo*) and the Caspian tern (*Hydropronge caspia*), the world's largest tern, both of which spend more time at sea compared with that of the gulls. On the other hand, waders or shorebirds spend all their feeding time scurrying along the flat beach just along the tidemark, darting here and there looking for worms, small crustaceans, and other arthropods. Examples of such waders are the western sandpiper (*Calidris mauri*), the smallest shorebird, the least sandpiper (*C. minutilla*), the dunlin (*Calidris alpine pacifica*), the lesser yellowlegs (*Tringa flavipes*), Wilson's snipe (*Gallinago delicata*), the black oystercatcher (*Haematopus bachmani*), the red-necked phalarope (*Phalaropus lobatus*), the killdeer (*Charadrius vociferus*), the black-bellied plover (*Pluvialis squatarola*), and the semipalmated plover (*Charadrius semipalmatus*) (Audubon Seattle, ND).

Boundary Bay is one of the largest undeveloped coastal wetlands, eelgrass beds, and salt marshes in northwestern Washington State and is home to many native birds as well as being an important stopover point on the Pacific Flyway for western sandpiper and dunlin; it has been designated a wildlife sanctuary by the Western Hemisphere Shorebird Reserve Network and is a Hemisphere Reserve (WHSRN 2008a, 2008b). One noted resident during the breeding season is the marbled murrelet (*Brachyramphus marmoratus*) an auk; they range from the Aleutian Islands south along the coastal regions as far as northern California, however it is an endangered species (BirdLife International, 2016). The semi-palmated plover (*Charadrius semipalmatus*), least sandpiper, red knot (*Calidris canutus roselaari*), the red-necked phalarope, a rail, the sora (*Porzana carolina*), and the black-bellied (or gray) plover (*Pluvialis squatarola*) are also common during migration. The refuge is also used by peregrine falcon, bald eagle, northern harrier, Caspian tern (*Hydroprogne caspia*), great blue heron, song-birds, and a variety of waterfowl (USFWS, 2021). Boundary Bay can expect more than 100,000 birds during the migrations (BBWMA, 2021). Other Hemisphere Reserves in the region are Birch Bay, 160 km north of Seattle, and Grays Harbor National Wildlife Refuge southwest of Puget Sound on the Pacific Coast. As many as 24 species of shorebirds use Grays Harbor National Wildlife Refuge, with the most abundant species being western sandpiper and dunlin.

SNAKES AND LIZARDS

Snakes and lizards (clade Reptilia) are amniotes that lay their eggs on land, the eggs having a gaseous-permeable, proteinaceous shell; unlike birds, they are exothermic and must warm under sunlight before they are able to hunt or forage.

Snakes

Let us begin with some of the snakes (Infraorder Serpentes) that inhabit the Puget Sound region. The relatively warm and dry climate, offered a multiplicity of habitats for small birds, mammals, reptiles, and invertebrates, the common prey for most snakes.

Garter snakes are some of the more common small (<1 m) snakes found almost everywhere in the Puget Sound region, each species specialized to a particular habitat (UWCAS, 2021). The Puget Sound garter snake (*Thamnophis sirtalis pickeringii*) is such an example. Their prey are generally red-legged frogs and juvenile bullfrogs. Interestingly, garter snakes are one of the few animals capable of ingesting the toxic rough-skinned newt (*Taricha granulosa*) without incurring sickness or death (UWCAS, 2021; Williams and Brodie, 2003).

The northwestern garter snake (*Thamnophis ordinoides*) is smallest of the garter snake species in Washington; it is found throughout the region along the coasts and has the most varied coloration within single populations. Those having a striped pattern evade predators by escaping in a straight line, whereas spotted or pattern-less snakes tend to dart in one direction, pause, then dart again, thereby using camouflage to escape (UWCAS, 2021). Some have red stripes along their body, thus leading some to misname them as "red racers" similar to their distant relative, the western racer (*Coluber constrictor*), which is found in eastern Washington state (UWCAS, 2021).

One of the more common snakes of the Puget Sound region is the northern rubber boa (*Charina bottae*); they are primarily nocturnal and feed upon the young of underground nesting rodents, such as shrews, voles, and mice, as well as lizard and snake eggs. The sale of wild-caught northern rubber snakes in the United States is illegal (UWCAS, 2021).

A less-common snake seen in the Puget Sound region is the common sharp-tailed snake (*Contia tenius*); their diet appears to be primarily gastropods, in particular, slugs and prefers to inhabit cool, moist regions. It reputedly uses its sharp tail to grip firmly its slug prey (UWCAS, 2021).

Another less-common snake is the Pacific gopher snake (*Pituophis melano-leucus*) and is found east of Puget Sound in hinterland and mountains of the Cascades (UWCAS, 2021).

Lizards

There are at least 15 species of native lizards (Order Squamata) that inhabit all the different ecosystems of the Pacific Northwest; there are also a few introduced species (UWCAS, 2021). Those native to the Puget Sound region include the alligator lizards (*Elgaria* spp.) and the northwestern fence lizard (*Scelopurs occidentalis occidentalis*). Introduced species include the Italian wall lizard (*Podarcis siculus*) introduced on Orcas Island since at least 1997, the common wall lizard (*Podarcis muralis*) introduced on Vancouver Island in 1970, and anoles (*Anolis* spp.) (California Herps, 2021).

The most commonly seen lizard throughout the Puget Sound region, usually sunning themselves on a rock, is the northern alligator lizard (*Elgaria coerulea*); they give birth in the summer to live young (UWCAS, 2021). The term "alligator" refers to the presence of supporting bony structures in their dorsal and ventral scales, as found in alligators. They can reach a size of over 25 cm including the tail (Stebbins, 2003).

TURTLES, TERRAPINS, AND TORTOISES

Turtles

A turtle native to Puget Sound is the northwestern pond turtle (*Actinemys marmorata*) (UWCAS, 2021). It is an endangered species in Washington and is only known from Puget Sound and the Columbia River basin (California Herps, 2021). The western painted turtle (*Chrysemys picta bellii*) is also seen in Puget Sound and the surrounding waterways; it is possibly being displaced in western Washington by the introduced red-eared slider (*Trachemys scripta elegans*) (UWCAS, 2021). Ironically, the western painted turtle is itself an introduced species in Southern California, as we noted in our companion volume "Making and Unmaking of the San Diego Bay" where it is most likely descendants of pets that had been abandoned by their owners (Kaser and Howard, 2021).

The two other turtles that may be seen in rivers, streams, and ponds in the Puget Sound region are both invasive. One is the eastern snapping turtle (*Chelydra serpentina serpentina*) which can be between 4.5 and 16 kg and which is native to ponds and shallow streams from the eastern United States to the Rocky Mountains, south through Mexico and Central America (UWCAS, 2021). The other is, as noted above, the red-eared slider (*Trachemys scripta elegans*) and whose original range were the rivers and waterways of the US mid-west to the south-western prairie, east to the Appalachians, and the Rio Grande valley in the south (California Herps, 2021).

AMPHIBIANS

Class Amphibia include salamanders, newts, frogs, and toads, as well as a limbless amphibian group, the caecilians. The largest of the Puget Sound region amphibians is the coastal giant salamander (*Dicamptodon tenebrosus*) up to 34 cm in length and is the largest terrestrial salamander in North America. It is endemic to the region and is found in humid forests and near mountain streams of the Puget Sound and the North Cascades (Frost, 2015).

A related salamander inhabits the Olympic Peninsula: Cope's giant salamander (*Dicamptodon copei*); it differs in appearance from the coastal species having a marbled gold and brown patterning and it attains a length of up to 19.5 cm (Hallock and McAllister, 2009)

Of particular interest, these are some of the many species of salamanders which can, under certain environmental conditions, retaining the larval form whist becoming sexually mature; this is termed "neoteny" and is considered in the tenets of biological sciences to be one way that evolution to adapt to changing conditions and speciation may occur, including in humans (Kucera, 1997; Gould, 1977). A neotenous salamander that may be familiar to the reader is the Mexican axolotl (*Ambystoma mexicanum*) and which has been frequently kept in aquariums as a pet; it frequently remains in the larval form throughout its life, which can be up to 25 years (California Herps, 2021).

A salamander frequently spotted in the damp woodlands in the hills and mountains of Washington and Oregon is a lungless salamander, the Oregon ensatina, *Ensatina eschscholtzii orɪgonensis*, and which has been the subject of continuous scientific study at the University of California at Berkeley since the 1940s. It was one of the first animals to be identified as a ring species, that is, the subspecies are distributed around a geographic feature and they eventually meet up at either end of the ring as two non-interbreeding species whereas those subspecies found in adjacent habitats "around the ring" could interbreed in what is termed a "hybrid zone" (Stebbins, 1959; Brown and Stebbins, 1964).

An example of another lungless salamander is the northwestern salamander (*Ambystoma gracile gracile*); it inhabits the coastal mesic regions of Puget Sound, the San Juan Islands, Cypress, Whidbey, Bainbridge, and Vashon Islands, and in the North Cascades to the tree line. It is notable from its long body (up to 22 cm including the tail) and also undergoes neoteny; it may require old-growth forests to thrive (Corn and Bury, 1991). Its coloration can vary from between very pale tan to sandy or ruddy brown. A related species, the long-toed salamander (*Ambystoma macrodactylum macrodactylum*) inhabits a wide range of ecosystems including temperate rainforests, coniferous forests, and riparian zones of the Olympic Peninsula, Puget Sound, and the western foothills of the North Cascades. It is characterized by its long forth toe on the hind limbs (Stebbins, 2003).

Originally believed to be restricted to the Columbia River Gorge, several specimens of the Larch Mountain salamander (*Plethodon larselli*) have been found in the eastern Puget Sound Basin, predominantly in the moist and lush talus slopes and old-growth forests; adults are typically less than 5 cm in length. The Center for Biological Diversity (Portland, Oregon) is petitioning for it to be listed as an endangered species (CBD, 2016).

The rough-skinned newt (*Taricha granulosa*) overlaps most of this range in the Puget Sound region, and is mainly found in the Pacific Northwest coast and mountain ranges of northern California, Oregon, Washington, British Columbia, and Alaska. Its skin exudes a neurotoxin, tetrodotoxin (TTX), which prevents the flow of sodium ions through voltage-gated sodium channels in the cell membrane of nerve cells and thereby blocking nerve impulses (Zakon, 2013). Interestingly, as mentioned earlier, one of its major predators, the common garter snake (*Thamnophis sirtalis*) exhibits resistance to the tetrodotoxin (Williams and Brodie, 2003).

The northern red-legged frog (*Rana aurora*) is found throughout the Puget Sound region, including Vancouver Island and the western flanks of the North Cascades (Stebbins, 1954). Although less common further south, it is not considered to be Threatened (IUCN, 2015).

The American bullfrog (*Lithobates catesbeianus*, but also known as *Rana catesbeiana*), is to be found throughout the Puget Sound region, predominantly in the coastal wetlands. Although native to North America east of the Rockies, is an invasive species to the western United States and was first sighted during the late 1890s. It preys upon birds, bats, rodents, frogs (even its own young), snakes, turtles, and lizards and thus may have deleterious ecological effects (McKercher and Gregoire, 2020). It prefers to live in permanently inundated wetlands and so constructive habitat modification by man, such as converting these wetlands to be ephemeral, may help to control the Puget Sound region's bullfrog population.

The Pacific treefrog (*Pseudacris regilla*), also known as the Pacific chorus frog, can either be brown or green in color, and can change from brown to green, depending upon the background brightness, in a matter of hours to days. Interestingly, its familiar call is almost always heard in movies as nighttime background, even in areas where it is absent in reality. Its diet mainly consists of invertebrates, often flying insects (California Herps, 2021).

The Western toad (*Anaxyrus boreas boreas*) is found throughout the Pacific coastal regions and western North America from southern Alaska in the north, to northern California, inhabiting washes, arroyos, sandy riverbanks, and other riparian areas. Its diet includes a wide variety of invertebrates, but it predominantly feeds on trail-forming tree ants (California Herps, 2021).

FISH

Fish consist of many classes, such as jawless fish (Agnatha; lampreys and hagfish), cartilaginous fish (Chondrychthyes; sharks and rays), and bony fish (Osteichthyes; ray-finned fish and lobe-finned fish); phylogenetic studies have demonstrated a more complex classification system, which we will not elaborate upon here.

Probably the most iconic fish of the Pacific Northwest is of course the Pacific salmon. Their image has been depicted on American Indigenous Peoples' art, clothing, canoes, totem poles, and in religious and cultural ceremonies for hundreds, if not thousands, of years (CRITFC, 2021). The salmon runs from the Pacific Ocean up the thousands of rivers and streams that are born in the western mountain ranges have been the source of nourishment for the many indigenous tribes who inhabited the coast and islands (Cannon, 2006).

The Chinook salmon (*Oncorhynchus tshawytscha*), also known as king salmon in Alaska, is the largest of the genus on the Pacific coast of North America, and is an anadromous (i.e., a fish that migrates up rivers from the sea to spawn) fish native to the Pacific Ocean and western North American river systems. It is the fish most spiritually and culturally prized amongst the Indigenous People (Cox, 1987), but their numbers have been falling since the 1990s, likely due to a combination of overfishing and the continued presence of dams along the many rivers where they had traditionally spawned (Behnke, 2010). Populations of Chinook are being restored into the Skagit River (Beamer et al., 2005). Records also show that even larger Chinook inhabited the western river systems before many of the hydroelectric dams were completed during the 1930s; known as "June hogs" they were up to 1.5 m in length and up to 55 kg in weight (Harrison, 2018).

The Coho salmon (*O. kisutch*) has also been the staple of many Indigenous People as well as being traded farther inland. In the Puget Sound/Strait of Georgia Evolutionary Significant Unit (ESU) it is a U.S. National Marine Fisheries Service (NMFS) Species of Concern (NOAA, ND).

The steelhead (*O. mykiss irideus*) is an anadromous coastal rainbow trout; the trout phenotype inhabits freshwater rivers and lakes, whereas its steelhead phenotype is a marine and ocean dweller. Compared to their cousins, the Chinook salmon (*O. tshawytscha*), they apparently spend fewer years in the freshwater waterways, probably due to of the inhospitable conditions often found in Washington streams (low flows, warmer temperatures) due to increased damming and have adapted to living under highly variable condition. They are also found in lagoons, which prove a more persistent and reliable water environment (Moyle et al., 2008).

Coastal cutthroat trout (*O. clarkii clarkii*) have similar life histories to the other anadromous salmonids in that although some remain in freshwater throughout their lives, many migrate from the ocean to overwinter and spawn in streams and rivers, then the juveniles migrate to the sea to feed and mature. Whereas steelhead and Pacific salmon (e.g., sockeye salmon [*O. nerka*]) migrate far out to sea, the cutthroat remain in or near estuaries, usually within 8 to 16 km of their spawning ground (Behnke and Tomelleri, 2002).

The peamouth chub (*Mylocheilus caurinus*) is a freshwater fish found throughout the Pacific Northwest; it is the only one of its genus and may grow up to 36 cm in length. It is a popular sports fish and is relatively toletrant to salt water and so it is found throughout the waterways of Puget Sound and the nearby islands of the Salish Sea (MFG, ND). They spawn in large numbers during May and June when water temperatures reach about 12 to 18° C.

Another freshwater fish is the northern pikeminnow (*Ptychocheilus oregonensis*) that can grow up to 63 cm in length and 3.5 kg in mass. They are of considerable concern to salmon hatcheries as they prey upon salmon smolt, that is, mature juveniles about to migrate to the ocean (Mesa, 1994).

The burbot (*Lota lota*) is a fresh water member of the cod family (Gadiformes) and is the only member of its genus. It too is a voracious predator and it preys upon lamprey, whitefish, grayling, young northern pike, suckers, stickleback, trout, and perch. They spawn between December and March, and often under ice where temperatures are only a couple of degrees above freezing (McPhail and Paragamian, 2000).

Pacific lamprey (*Entosphemus tridentatus*) has generally the same distribution in the Puget Sound region as the southern steelhead and are also anadromous; they are distinct from the non-parasitic western brook lamprey (*Lampetra pacifica*) which is common in the freshwater coastal waterways of Vancouver Island and Puget Sound (Vladykov and Follett, 1965; Smith et al., 1993). A related parasitic variety, the Morrison Creek lamprey (*Lampetra richardsoni var. marifuga*), is unique to Morrison Creek on Vancouver Island, and is listed as Endangered; its ammocoete (larval) phase of life history may play a major role in stream nutrient recycling. Within three to seven years, they undergo metamorphosis and become sexually mature usually within six to nine months; although they develop teeth, these are not maintained and become blunt and peg-like, and the adults no longer feed. Interestingly, some of the ammocoetes metamorphose into a *marifuga* form, whereby they retain sharp teeth for feeding and sexual maturation is delayed for at least a year (COSEWIC, 2010). Such changes in metamorphosal patterns within a single species are of fundamental interest to developmental biologists as they can provide models of how the endocrine system regulates such processes in an animal's life history.

Many other fish have been introduced into lakes and rivers for either recreational fishing (such as bass or catfish) or as bait (golden shiner or goldfish, for example); for reasons of brevity, we have not included a summary here.

Invertebrates

The word "invertebrate" is no longer used by biologists as a valid classification term and the organisms we will be describing below in fact are in many ways even more dissimilar to one another than mammals are to fish. The more correct terminology for these "animals without a backbone" organisms is "protostome," which describes a particular sequence of gut formation during embryonic development and which we need not describe in more detail here. In contrast, the echinoderms and the vertebrates or chordates are referred to as "deuterostome" (Martín-Durán et al., 2016). Therefore, although echinoderms are not protostomes, we will nevertheless include them in this section.

By almost any measure, the invertebrates are a fascinating group. In sheer numbers, there are more of them than any other animal group. They also have the greatest diversity of any group: they make up 95% of all animal species. Yet they are easily overlooked. They include insects, spiders, crabs, snails, clams, squids, octopuses, earthworms, leeches, jellyfishes, sea anemones, and many, many more. Most of these are represented in the Puget Sound region.

Invertebrates, whether aquatic, marine, or terrestrial, are important and often under-studied components of any habitat (Maffe, 2000). Few reports focus on them. Yet they serve as sensitive indicators of the overall health of the environment. Many invertebrates are important for other organisms. They act as pollinators, herbivores, scavengers, predators, and prey. Without the pollination of food crops by honeybees, human diets would be quite different. Unfortunately, the Puget Sound estuary are some of those areas in which these important organisms have been under-studied.

We have arbitrarily separated them into two groups: terrestrial and marine invertebrates.

Terrestrial Invertebrates

Insects

The Seattle-Tacoma-Bellvue (STB) metropolitan area is a major urban area with nearly four million people (www.statista.com). It is also a hot spot for insect diversity. The intersection of those sets of organisms also means that the Puget Sound region is a hot spot for threatened species (Dobson et al., 1997).

The insect diversity in the Puget Sound region results from two factors. First, Washington state overall is a biological hotspot due to its geology, soil, and climate. Its latitude ensures a warm and humid Mediterranean climate, and the elevation varies from sea level to the mountains of the Olympic Peninsula and the North Cascades. Second, the combination of the hills, estuaries, and bays yields a multitude of microenvironments with different climates and soils.

Unfortunately, much of the STB region has been developed. Buildings and streets cover more than 40% of the developed land, and the habitat for insects has been lost. For example, about 43% of the butterfly species have been lost. In addition to loss of habitat, those areas that can support insects have been fragmented. As is the case with many species of wild animals, the inability to move from one area to another is highly stressful. Also, the quality of those undeveloped areas does not allow them to support a sufficient population of some insects. Invasive species have also put pressure on native insects. These might be insects that compete with the native insects or even invasive plant species that crowd out native plants and reduce food sources for insects. Ice plant (*Carpobrotus edulis*), which is frequently used as easy to grow and low-maintenance groundcover, changes the nature of the soil that it grows in so that there are fewer insects and other invertebrates.

Preventing exotic insects from gaining a foothold is the best strategy for dealing with invaders. However, it is not always possible, and so, additional strategies are necessary to eradicate non-native insects (Liebhold and Kean, 2019). Many nascent infestations fail because the insects cannot establish themselves in the new environment. They might not find suitable food or habitat or there might be too few insects to succeed. However, once the non-native insects get a start, controlling them in a forest is very challenging. The good news is that not every insect must be killed. The population only needs to be reduced below a threshold.

The Burke Museum of Natural History at the University of Washington has an excellent website that lists and describes many of the insects of the Puget Sound region (https://www.burkemuseum.org/). Some insects are more welcome than others by the general public: bees, butterflies, and dragonflies.

Insects (subphylum Hexapoda, having six legs) are classified into several groups: Collembola (springtails), Protura (coneheads), Diplura (two-pronged bristletails), Archaeognatha (jumping bristletails), Zygentoma (silverfish), and Pterygota (winged insects) (Chinery, 1993). In this chapter, we will focus on the Pterygota since although wingless insects are important herbivores and detrivores in the ecosystem, their presence is usually undetected under most contemporary circumstances, and winged insects are those that the reader is most likely to encounter.

Dragonflies and damselflies (order Odonta) are a more primitive insect; their ancestors first appearing during the Carboniferous Period, around 325 mya (Resh and Cardé, 2009). They have two pairs of wings rather than the single pair of wings and the paired halteres or the paired elytras of more advanced flying insects (e.g., flies or beetles). Some of the common dragonflies include common green darner (*Anax junius*), blue-eyed darner (*Rhionaeschna multicolor*), wandering glider (*Pantala flavescens*), American bluet damselfly (*Enallagma* spp.), and, of particular note, the dragonhunter (*Hagenius brevistylus*), a type of clubtail dragonfly that hunts smaller dragonflies and other insects (II, 2021).

Grasshoppers and crickets (order Orthoptera) have been around for 250 million years. These ground-dwelling insects have powerful rear legs that allow them to escape predators. Those in the Puget Sound region include the painted grasshopper (*Dactylotum bicolor*), the bird grasshopper (*Schistocerca* spp.), which when swarming, are called locusts, and the house cricket (*Acheta domesticus*).

About 180 species of butterflies and moths (order Lepidoptera) are found in the Puget Sound region. The monarch butterflies (*Danaus plexippus plexippus*) migrate through the area on their way to and from their summer breeding areas in Canada; they and their larvae feed almost exclusively on milkweed (*Asclepias linaria*). Milkweed leaves and nectar contains cardenolides, steroid glycocides, which are generally cardio-toxic to animals, but not to the monarch; these compounds build up in the caterpillar's haemolymph (body fluid) (Agrawal, 2017). Their orange and black wing patterning indicates that they are foul-tasting, thus evading predators. A butterfly which mimics the wing pattern of the monarch is the viceroy (*Limentitis archippus*); its larvae feed on plants of the willow family that contain salicylic acid (a chemical precursor of aspirin), so its bitter taste would also be unpalatable to a predator; this is termed Mullerian mimicry, named after Fritz Müller, a 19th century German zoologist who first described the phenomenon in insects in the Brazilian forests (Müller, 1878). Other butterfly species of the Puget Sound region include the western tiger swallowtail (*Papilio rutulus*), western pygmy blue (*Brephidium exilis*), Lorquin's admiral (*Limenitis lorquini*), and painted lady (*Vanessa cardui*). Also seen are the two-tailed swallowtail (*Papilio multiculdata*), the great spangled fritillary (*Speyeria cybele*), and Milbert's tortoiseshell (*Nymphalis milberti subpallida*). The hummingbird moth (*Hemaris* spp.) is notable in that it feeds both during the day as well as at night (II, 2021). An unusual moth is the dogwood borer (*Synanthedon scitula*); its body shape and coloration mimics that of wasps and its larvae (caterpillars) bore their way into tree bark and feed on the cambium layer of the tree; they are considered pests by arborists and horti-culturalists (II, 2021). The purplish copper (*Lycaena heloides*) and the cedar hairstreak (*Callophyrus gryneus*) are indigenous to the Puget Sound region.

Beetles are in the order Coleoptera. They are different from most other insects in that their front wings have hardened into wing cases called elytra. With 400,000 species, beetles are an amazingly successful and diverse group of or-ganisms. They account for about 40% of all insects and 25% of all known an-imals. They are well-represented in the Puget Sound region with many species, including the convergent lady beetle (*Hippodamia convergens*), the mountain pine beetle (*Dendroctonus ponderosae*), and the northern carrion beetle (*Thanatophilus lapponicus*). An invasive species, the Asian multicolored lady beetle (*Harmonia axyridis*) is native to China, Japan, and eastern Siberia and was probably introduced in the late 1970s; it is believed to be displacing the native convergent lady beetle (II, 2021).

Hemiptera are often referred to as the "true bugs" in that they all possess distinctive mouthparts adapted to suck fluids from an organism, whether it be a plant or a person (Ruppert et al., 2004). They include aphids (*Aphis* spp.), scale insects, such as San Jose scale (*Quadraspidiotus perniciosus*), shield bugs (*Elasmostethus cruciatus*), assassin bugs (*Pselliopus* spp.), bed bugs (*Cimex lectularius*), stink bugs (*Halyomorpha halys*), as well as cicadas (*Platypedia areolata*) and exotic leafhoppers (*Japananus hyalinus*) (II, 2021).

The order Hymenoptera includes, ants, bees, wasps, wood wasps, and saw-flies; many of these are the social insects. The Puget Sound region has about 170 species of native bees from six families (Apidae, Andrenidae, Colletidae, Halictidae, Megachillidae, and Melittidae). Honeybees (*Apis mellifera*) are an introduced species, originally from Europe. Other common bees include the yellow bumblebee (*Bombus fervidus*), yellow-faced bumblebee (*B.*

*vosnesenskii*or *B. caliginosus*), the small carpenter bee (*Ceratina* spp.), and the digger bee (*Habropoda* spp.). The European wool carder bee (*Anthidium manicatum*) is an introduced species that is spreading rapidly.

Wasps are narrow-waisted insects that can sting. There are tens of thousands of species. While some live in colonies, most are solitary. They are valuable pollinators. Several species occur in the Puget Sound region, including the common paper wasp (*Polistes exclamans*) and western yellowjacket (Vespula pensylvanica). A common parasitic wasp is the giant ichneumon wasp (*Megarhyssa nortoni*), which although startling in its size, bears a long ovipositor, equal in length to its own body, which it uses to bore into tree bark under which lies the eggs or larvae of the horntail wasp (*Urocerus* spp.) which are there to feed upon the cambium. When the parasitic larvae emerges from the egg, it consumes the horntail larvae until it metamorphoses, burrows out of the bark, and is then ready to mate (II, 2021). One wasp to be watchful for is the so-called eastern velvet ant (*Dasymutilla occidentalis*); it is in fact a wasp, mimicking an ant, but its wingless female can inflict an extremely nasty sting if handled (II, 2021). They parasitize bumble bees and lay their eggs in a bumble bee nest; once the eggs hatch, the larvae feed on the bumble bee larvae. It is also known as the cow killer, due to its sting.

The Puget Sound region used to be home to over 100 species of ants. Of note, the introduced Argentine ant (*Linepithema humile*), which has driven out most of the native ants in urban areas in much of coastal California does not appear to have migrated north of its supercolony; the reasons to date are uncertain but may be due to a variety of factors, including access to water, winter temperatures, and lack of suitable physical access to the Puget Sound area (mountain ranges) (Markin, 1969). Other native ant species include the carpenter ant (*Camponotus modoc*), the pavement ant (*Tetramorium caespitum*), the odorous house ant (or sugar ant) (*Tapinoma sessile*), the western harvester ant (*Pogonomyrmex occidentalis*), and the western thatching ant (*Formica rufa*).

The order Diptera includes all the flies, such as houseflies (Muscoidea), horseflies (Tabanomorpha), blowflies (Oestroidea), crane flies (also called "daddy-long-legs") (Tipuloidea), hoverflies (Syrphoidea), mosquitos (Culicomorpha), tsetse flies (Hippoboscoidea), and screwworms (larvae of blowflies). They are major predators, pollinators, and detrivores; many of their larvae are important agricultural and domesticated animal pests. They are characterized by having only a single pair of wings, the second pair have been altered during metamorphosis in the maggot into a pair of small club-shaped halters, organs used to fine-tune balance during flight.

These are only a few of the insects in the Puget Sound region. There are also praying mantises, walking sticks, and, unfortunately, lice. However, in recent years, scientists have become worried about the decline in the numbers of insects. The cause of the decline seems to be human activities (Simmons, 2019). A series of papers has sounded the alarm. The loss of insects will affect our food supply. Hallmann et al. (2017) used traps across Germany to determine the number and types of insects. They found that flying insects have decreased by 75% over the 27 years of study. Lister and Garcia (2018) documented a similar decline in arthropods in the soil. Mathiasson and Rehan (2019) showed a loss of a significant number of bee species. Beekeepers noted that they lost 40% of their hives in the winter of 2018–2019 (Neilson, 2019). The causes are thought to include decreasing crop diversity, poor beekeeping practices, loss of habitat, and indiscriminate use of pesticides. Infestation by a mite *(Varroa destructor)* has also harmed hives.

Arachnids

Arachnids are a large class of invertebrates that feature spiders, scorpions, ticks, mites, harvestmen, and more. Most have eight legs, but some have other appendages that might look like legs. In addition, there are some mites that have fewer legs. Unlike insects, they have no antennae or wings, and have two main body segments rather than the three of insects.

Spiders

Spiders are very common, essentially everywhere, and the Puget Sound region has its share. According to the website "SpiderID" (spiderid.com), there are 34 species observed in the state of Washington from confirmed sightings (SpiderID, 2018). The rabbit hutch spider (*Steatoda bipunctata*) has occasionally been seen in Washington, usually in the late spring; it appears to be native to northern Europe and North America but is rarely seen south of the US-Canada border (SpiderID, 2018). In appearance it resembles that of the western black widow (*Latrodectus hesperus*) but its bite is somewhat less severe (Platnick, 2010). A more common spider is the long-bodied cellar spider (*Pholcus phalangioides*); it is gray colored with an elongated abdomen and is mostly spotted indoors. The Johnson jumping spider (*Phidippus johnsoni*) appears to have a more restricted range, confined to western North America; it is mainly seen outdoors and rarely in a web (SpiderID, 2018). It has a distinctive scarlet markings along the upper sides of its abdomen. The common candy-striped spider (*Enoplognatha ovata*) is native throughout Europe and is introduced to North America (mainly Washington and Montana); it is a ferociously small (not > 6 mm) predator and preys on insects many times its size (Preston-Mafham, 1998).

Of the most notorious venmous spiders in North America, the western black widow (*Latrodectus hesperus*) and the American yellow sac spider (*Cheiracanthium inclusum*) are frequently found in forested areas of the Puget Sound region (WSDOH, ND).

Crab spiders (*Misumena* spp.) are very small spiders that do not spin webs; instead they lie in wait for their prey, nectar-feeding insects and pollinators, within a petal cluster of a flower and are camouflaged by controlling the color of their cuticle. They are able to change from yellow to pure white (as seen in normal daylight) over a period of only a few hours. Under ultraviolet light, however, their yellow pattern is invisible to an insect, whereas their white pattern actually stands out under UV light but is mistaken as part of the plant's structure by the incoming insect (Morse, 2007). In the Puget Sound region, the native is the golden-rod crab spider (*Misumena vatia*) (SpiderID, 2018).

Scorpions

The Puget Sound region has very few scorpions (Buhler, 2018). They are rarely seen as they are nocturnal. They are also not dangerous unless the person stung happens to be allergic to the venom. Amazingly, they can most easily be found by using a black or ultraviolet light at night. The scorpions fluoresce brightly. They give birth to live young that ride on the mother's back early on. The most common species in the Puget Sound region is the northern scorpion (*Vejovis boreus*); they can be found on dry southwestern slopes (WSU, ND).

Ticks

A tick of concern that inhabits the Puget Sound region is the western blacked-legged tick (*Ixodes pacificus*); it is a carrier of Lyme Disease (*Borrelia burgdorferi*), as well as another bacteria that can cause anaplasmosis, and the parasitic protist *Babesia microti* which results in babesiosis, a malaria-like disorder, infecting red blood cells. Hikers in the region are encouraged to shower off and clean their clothing after an outing (CDC, 2021).

ISOPODS (WOODLICE)

These small crustaceans have segmented, dorso-ventrally flattened bodies with seven pairs of jointed legs. Females carry the fertilized eggs in their marsupium on the ventral side of her abdomen. The young emerge and seem to be a live birth, but they are really from eggs. There are thought to be 5000–7000 species worldwide. Most are useful in that they turn the soil like earthworms. However, they do eat some crops (e.g., strawberries) and are considered a pest if they invade homes. The most common isopod, from the suborder Oniscidea, reported from western Washington and the Puget Sound region is called the rough porcellio or scabby sowbug (*Porcellio scaber*); it was reported as having invaded greenhouses as a pest and it may have been introduced (Hatch, 1947).

One other common group is the pill bugs or rolly-pollies (*Armadillidium vulgare*) that are familiar to most children. They are common under rocks, leaves, or boards in the backyard. They were introduced to the Americas from Europe, where they are known as woodlice, but have thrived here. They are the most successful group of terrestrial crustaceans (Ruppert et al., 2004). They breathe through gill-like slits and so require moisture to survive.

Interestingly, the superorder to which they belong, the Peracarida, is one of the largest crustacean taxa and includes about 12,000 species; they originally inhabited, and many still do, marine environments and may have been amongst the first true land animals (Poore, 2002).

MYRIAPODS (MILLIPEDES AND CENTIPEDES)

At first glance, millipedes and centipedes look like insects with lots of legs. However, they are quite different. Millipedes have two pairs of legs per body segment. While a small number are predators, the vast majority eat decaying organic matter, detrivores. Centipedes on the other hand have only one pair of legs per segment and generally have fewer segments. They are essentially carnivorous. They were also some of the earliest arthropods to colonize land during the Carboniferous Period (320 to 290 mya), the largest being *Arthroplura*, which could be at least 2 m in length and 50 cm in width.

There are a number of species of each myriapod found in the Puget Sound region (Shelley, 2002). The brown millipede *Tylobulus claremontus* is about 3–9 cm when mature. It is black to dull reddish-brown. They are valuable to the ecosystem because they break down plant matter (especially conifer litter) into humus. Leschi's millipede (*Leschius mcallisteri*), which was only discovered and named as a new species in 2004, is considered under the Washington State Wildlife Action Plan as under threat and in need of more research (WDFW, 2015). The greenhouse millipede (*Oxidus gracillis*) is an introduced species almost globally in both the tropics and temperate zones. It is about 2 cm in length and is often considered a pest in greenhouses.

There are four groups of centipedes living in the Puget Sound region. Stone or brown centipedes (*Lithobius* spp.) are small (usually between 1.5 and 3 cm) and have 15 pairs of legs as adults. They are common in gardens and eat small insects. Soil centipedes (*Strigamia bidens*) can be quite large (up to 195 mm) and can have more than 60 pairs of legs. They live in the ground and eat subterranean insects. House centipedes (*Scutigeromorpha coleoptrata*) include only one species. They have 15 pairs of long legs and are about 5 cm long. These are the only local centipedes that can bite, but their fangs are weak. The cryptoid centipede (*Theatops californiensis*) is restricted to western North America including Puget Sound and often invades homes. They are one of North America's smallest centipede, reaching a length of between ten to 65 mm; although small, it often goes unnoticed and its bite to bare feet can induce a painful reaction. Of great concern is the blue-gray taildropper centipede (*Arctogeophilus insularis*) which is listed by the Washington DFW as imperiled (WDFW, 2015).

Gastropods (Snails and Slugs)

Snails and slugs are both mollusks. Slugs have lost the shell that covers snails. All produce mucus to facilitate their movement and protect against desiccation. For a more extensive list, the reader is referred to the *A Field Guide to Some Snails & Slugs in Washington for Survey Inspectors* (Marquez, 2013).

Slugs and snails have a radula that they use for feeding. By eating leaves, animal droppings, moss, and dead plant material, they produce soil humus and provide a valuable service. Two common species in the Puget Sound region are the three-banded garden slug (*Lehmannia valentiana*) and the gray garden slug (*Deroceras reticulatum*) (Winters, 2016). Both are terrestrial slugs and feed on plants and decomposing wood; they are also considered pests as they enjoy orchids and ornamental plants (South, 1992; Baker, 1999)

The most common snail in the Puget Sound region is the non-native European garden snail (*Cornu aspersum*), but there are several native species (UCANR, 2021). The native Pacific sideband snail *(Monadenia fidelis)* has a very dark shell and beautiful purple flesh. They are the most common native land snails in the Puget Sound region and are easily confused with garden snails. They are found mostly in undisturbed native ecosystems.

The non-native leopard slug (*Limax maximus*) are quite aggressive and will attack other snails as prey, including the Pacific banana slug (*Ariolimax columbianus*) and the taildropper slug (*Prophyaon andersoni*) (Sandelin, 2007). Banana slugs are so named because of their general shape and bright yellow coloration; it is thought by some that their yellow colors match the fallen, autumnal bay leaves common to some of their habitats, which renders them camouflaged to birds and small mammals. However, many of them may also be green, brown, tan, and white, and they can grow to a length of nearly 25 cm. By eating leaves, animal droppings, moss, and dead plant material, they produce soil humus and provide a valuable service. They are the second largest land slug in the world (Burke, 2013).

The robust lancetooth *(Haplotrema vancouverense)*, another slug, eats plant material, and other snails and slugs (Winters, 2016).

Worms

We are most familiar with earthworms, but there are many, many others. Earthworms are not well studied in the Puget Sound region and many of those here have been introduced. Native earthworms can still be found in undisturbed areas. Vancouver Island and the Olympic Peninsula have at least nine native species in three genera (*Bimastos lawrenceae, Arctiostrotus vancouverensis, A. adunatus, A. pluvialis, A. fontinalis, A. johnsoni, A. perrieri, A. altmani,* and *Toutellus orognensis*) (McKey-Fender et al., 1994). Non-native worms of the families Acanthodrilidae, Lumbricidae (comprising the familiar European earthworm [*Lumbricus terrestris*] and its kin), Megascolecidae, Ocnerodrilidae, and Sparganophilidae may also be found. The rest are invaders. Reynolds (2016) has an extensive list of all of the species and their locations.

Many worms are parasites, and some parasitize humans. Nematodes, which are roundworms, occur in many species of fish. They appear on the intestines, liver, in the body cavity or in the flesh. They also live in seals, porpoises, whales, and dolphins. Their eggs escape in the feces of the mammal and are eaten by small crustaceans that are then eaten by fish or squid. The nematodes can also infect humans if they eat raw or unprocessed fish. Flukes and tapeworms are also common in the Puget Sound region fish and other animals.

MARINE INVERTEBRATES

Many invertebrates are found in the Puget Sound, the Salish Sea, and the Pacific Ocean (NPS, 2018). The numbers of different types of organisms are almost beyond counting. They can be best seen while snorkeling, scuba diving, or just looking at tide poolsand they are easily visible at low tide.

Here we will name a sampling of common species. For a more detailed account of marine invertebrates, we recommend this outstanding book *Intertidal Invertebrates of California* by Morris et al. (1980).

PORIFERA

Sponges

The bodies of sponges have pores that allow water to circulate. They are filter-feeders and feed on single-celled organisms, detritus, and other material that they filter out from the water. The Georgia Strait is home to one of the only reefs made up of glass sponges in the world and which had been thought to be extinct for about 100 million years. The other sponge reefs are off the coasts of Washington and British Columbia; one, in the Hecate Strait, is up to 25 m in height, covers an area of about 60 km^2, and is at least 9000 years old (Jones, 2015). Glass sponges (class Hexactinellida) require silica to make their skeletons and it so happens that the Pacific Northwest coastal ranges comprise silica-rich feldspar granites; these are then weathered and the dissolved minerals are washed out to the bays by the numerous rivers (Hines, 2007; Jones, 2015).

There are more than 265 sponges documented in the waters of Puget Sound, including the cloud sponge *(Aphrocallistes vastus*, a glass sponge), the hard gnarled clump sponge *(Xestospongia hispida)*, the peach ball sponge *(Suberites montiniger)*, the orange puffball sponge *(Tethya aurantium)*, and the glove sponge *(Neoesperiopsis digitata)*; the latter are of the class Desmospongiae, having calcium carbonate as the skeleton-forming mineral, with silica spicules (Collins, 2020).

The red beard sponge *(Clathria prolifera)* is a red or orange-brown sponge with many projections. This sponge was originally from the Atlantic Ocean and arrived in Willapa Bay in the 1960s (Cohen, 2011). The yellow sponge *(Halichondria bowerbanki)* appears as a flat mass on objects. It eats plankton. It was also from the Atlantic, but arrived in the Pacific Northwest in the 1950s.

CNIDARIA

The phylum Cnidaria also includes jellyfish (jellies), sea anemones, corals, sea pens, and gorgonians.

Jellies

Jellies are graceful, beautiful, and sometimes dangerous. They go with the flow, literally. The current moves them along, but a few can also propel themselves to a degree. They vary in size from almost microscopic to specimens with tentacles of over 35 m. Among the many species found in the Puget Sound region are the following. The moon jelly *(Aurelia labiata)* resembles a large dinner plate;

aggregations of this jelly can be so large that they can be seen from aircraft flying over Puget Sound and may number in the millions (Norris, ND). It is translucent to whitish with short tentacles at the rim and four longer tentacles at the center. It eats plankton that is captured in the mucus on the tentacles and passed up to the mouth in the bell. The lion's mane jellyfish (*Cyanea capillata*) is probably the largest jelly in the world, having a bell diameter of up to 2 m and with tentacles up to 30 m in lengh. It eats zooplankton, which are transported to the mouth as in the moon jelly and is not tolerant to warmer waters of the lower lattitudes. Unlike the better-known, but generally smaller, Portuguese man o' war (*Physalia physalis*, a colonial syphonophore), it matures into the adult form from a single medusa. The comb jelly (*Pleurobrachia bachei*) is transparent with eight rows of "combs" or cilia. It has two tentacles that collect food. These jellies bioluminesce in the dark. Their cilia aid in movement (EOPS, 2021a).

Sea Anemones

Anemones look like underwater flowers (Gong, 2019). They have two lives: first, they form stationary polyps attached to a substrate, and second, they produce eggs that become mobile planula larva. All anemones are carnivorous. The giant plumose anemone (*Metridium farcimen* [*giganteum*]) can grow to over 1 m in length and up to 10–11 cm wide and has more than 100 tentacles on its lobbed oral disk. The common Washington aggregating anemone (*Anthopleura elegantissima*) eats anything that gets caught in its stinging tentacles (Jamison, ND). Another anemone familiar to tide-poolers is the giant green anemone (*Anthopleura xanthogrammica*); it is the oval-shapped green anemone of about 15–20 cm in diameter having multiple rows of conical stinging tentacles. The orange-striped green anemone (*Diadumene lineata*) attaches to various underwater objects. It is shiny green, olive, or olive brown with orange stripes. This non-native anemone was introduced years ago to the Pacific Northwest from the Pacific coasts of Asia and Japan, and it is now found worldwide (Carlton, 1979).

Sea Pens

Sea pens are classified as octocorals, also termed "soft coral", and are a third type of cnidarian. In the Puget Sound region, two cnidarian species are of note, the orange sea pen (*Ptilosarcus gurneyi*) and the slender sea pen (*Stylatula elongata*). They are made up of a colony of polyps each of which may differentiate into a different part of the organisms' anatomy—some may form the anchoring base while others become the animal's rachis, the feeding stalk. In addition, the polyps are further classified into (a) autozooids, which wave tentacles to catch prey, drifting plankton, and (b) siphonozooids, that distribute any food taken by the tentacles and distribute food throughout the colony using a water circulation system, perhaps a primitive ancestor of the vertebrate circulation and/or lymphatic systems. The autozooid may have nematocysts that are individual structures comprising a poison-tipped dart-like structure present at their tips, capable of paralyzing small animal prey. In addition, the slender sea pen autozooid conceals needle-like spikes (sclerites) which further discourage predation. When touched, both the orange and slender sea pens emit a greenish-blue light (bio-luminescence), probably as a defense, warning, or to startle—these cells can then be ingested by predators (Burgess and Eagleston, 2016a, 2017; Francis and Sire de Vilar, 2020). Upon nervous stimulation, the light is generated biochemically by endogenous luciferin and luciferase to emit blue light, which further interacts with green fluorescent protein (GFP) to create the blue-green light (Ward and Cormier, 1978; Francis and Sire de Vilar, 2020).

ANNELIDS

Flatworms

Flatworms belong to the phylum Platyhelminthes (Watkins, ND). Two of the three classes Cestoda (tapeworms) and Trematoda (flukes) are parasitic, and one, Turbellaria, is free-living. They are very simple and lack a respiratory or circulatory system. They absorb oxygen through their skin. Their mouth is at mid-body. Food enters the mouth and waste is expelled from the mouth also. The free-living flatworm *Notoplana acticola* is 2.5–7.6 cm long with two eyespots on the dorsal surface. It is brown or gray and eats small crustaceans, zooplankton, other worms and dead animals. Flatworms are very colorful and graceful in the water.

Round Worms and Worms

Round worms belong to the phylum Annelida and they are both terrestrial and marine. The Japanese green syllid (*Megasyllis nipponica*) is, eponymously, originally from Japan (Tighe, 2019). They live mostly in the mud, rocks, and other places at the bottom of the Sound. In the summer, they change radically. Adult worms move eggs and sperm to the rear of the worm and bud off a new worm. This schizogamy yields bright orange epitokes that swim around looking for partners. Upon finding a partner, they burst open to allow the sperm to fertilize the eggs. The native fifteen-scaled worm (*Halosydna imbricata*) is about 6 cm long with 18 pairs of dorsal scales. It can be red, tan or brown. They are scavengers. Some live in the tubes of tubeworms in a form of commensalism.

The leech *Branchellion lobata* is about an inch or so long with a sucker at each end. It feeds by attaching to sharks, fish and squids and drinking their blood. They begin life as males and later become females.

Like many invertebrates, the worms can also be quite strange. The scolecidan *Abarenicola pacifica* (the Pacific lugworm) inhabits a J-shaped burrow in both intertidal and subtidal mudflats. It "plugs" the burrow and uses its body to pump seawater through the burrow so that it can capture food (Arp et al., 1992). Other invertebrates live with the worm and feast on its leavings (Parr, 2019). These include the clam (*Cryptomya californica*), the scale-worm (*Hesperonoe com-planata*), and the Schmitt pea crab (*Pinnixa schmitti*). Strong storms can break up the sand in which the worms live, and occasionally, thousands of them can be seen stranded on beaches of Puget Sound.

MOLLUSKS

Many mollusks are sessile or creep slowly across a surface on their (usually) single foot; however, in the larval stage, they are free-swimming trochophores and which make up a large proportion of the zooplankton in the ocean surface. We will describe a number of mollusks that are typically found on the Puget Sound shores.

Chitons

Chitons are small- to medium-sized mollusks having a single upper shell, made up of about eight separate segments, which is both flexible (for accessing odd-shaped spaces) and hard (for protection from predators). In Washington, the largest found, up to 36 cm in diameter, is the gumboot chiton, *Cryptochiton stelleri*, and this inhabits the cooler Pacific waters north of the central coast. In addition, the smaller Katy chiton, *Katharina tunicata* (10–12 cm) is to be found in deeper waters, often up to 40 m deep (Lunsford and Helmstetler, 2003).

Clams

Clams are bivalves. Their two calcerous shells are connected by two adductor muscles, and they have a power foot for burrowing into the sand. Unlike oysters and mussels, they do not attach to a substrate. The term "clam" is typically used to describe those bivalves that are edible. They are all filter feeders.

There many species of clam that inhabit the Puget Sound seashore. The Washington Department of Fish and Wildlife notes that the Pacific razor clam (*Siliqua patula*) is one of the most sought-after shellfish in the state of Washington (WDFW, 2021b). They are to be found buried in the sand at the tideline and in the fall and winter, during night-time low tides, thousands of people can be found on the shore digging for the clam. A clam license is required (WDFW, 2021b) and it is one of the most commonly found on the beaches.

The residents of Puget Sound have probably been harvesting the huge calm, the geoduck (*Panopea generosa*), for thousands of years; it is one of the largest burrowing clams in the world with an up to 20 cm shell and a siphon that can extend to almost a meter in length. It too has now been the subject of restricted taking since 1925 and a person may only take three clams per day (Williams, 2021).

The boring softshell clam (*Platyodon cancellatus*) is white or gray with a dark siphon and the small California sunset clam (*Gari californica*) is white with pink lines along the growth zones (iNaturalist, ND).

Invertebrates, including clams, can also be very destructive. For centuries, the shipworm *Teredo navalis* was the bane of wooden ships and other wooden structures in the water. Although it is called a "worm," it is actually a clam.

Mussels

The bay mussel *Mytilus trossulusus* and the western ridged mussel (*Gonidea angulata*) are also bivalves; both are bluish gray. The latter is listed as

Endangered under the Canadian Species at Risk Act. These filter-feeders eat detritus and microscopic plants and animals. They form dense clusters on rocks, piers, and other substrates that are in calm waters. Researchers have recently discovered that the bay mussel often hybridizes with the invasive Mediterranean blue mussel (*Mytilus galloprovincialis*) (Elliott et al., 2008).

Oysters

Oysters are bivalve filter-feeders that eat plankton, bacteria, and detritus. They are found throughout the intertidal zone. The Olympia oyster (*Ostrea lurida*) is the common native oyster in the Puget Sound region. The shell is 5–8 cm long with a wavy edge. Recently the Pacific oyster (*Magallana gigas*), native to Japan but cultured locally, has become an invasive species; it has an important effect upon the environment as it builds reefs and this may change the shoreline ecosystem. It is also a threat to the native oyster restoration program (Johnson, 2013).

Oysters have been important in the Pacific Northwest for hundreds of years. The Native Americans used them as a handy source of food and built mounds with the left-over shells.

Marine Snails

Marine gastropods are some of the more common mollusks seen in the Puget Sound region. Common marine snails include the sea marsh periwinkle (*Littorina subrotundata*), which inhabits the salt marshes and the splash zone at high tide; the California assiminae (*Angustassiminea californica*), also a native salt marsh inhabitant, the European melampus (*Myosotella myosotis*), another salt marsh inhabitant but introduced from Europe, and the Manchurian cecina (*Cecina manchurica*), a species introduced from Japan and which lives in high intertidal brackish water (Burke, 2013).

Nudibranchs

The words "sea slugs" might conjure up a sea-going version of the slugs we see on land. Nothing could be further from the truth. Sea slugs are carnivorous marine gastropods and come in many colors; many are festooned with cerata (soft projections) (Ueda and Agarwal, ND). As a result, common nudibranchs, sea hares, sapsucking slugs are some of the most beautiful of sea creatures. Over 3000 species of nudibranchs are known but there are only about 20 species that inhabit Puget Sound (Bergamin, 2014; Burgess and Eagleston, 2016b). As larvae, they possess a shell but this is shed when they mature as adults (Thompson, 2009).

There are two forms of nudibranchs. Dorids are large, round, and flat, having a posterior branchial plume functioning as a gill. Aeolids are smaller with lush, feathery digestive and respiratory structures called cerata. One of the most common nudibranchs native to the Puget Sound region is the striped nudibranch (*Armina californica*); they feed mainly upon two cnidarian species, the orange sea pen (*Ptilosarcus gurneyi*) and the slender sea pen (*Stylatula elongata*). As noted earlier, these sea pens exhibit bioluminescence in their cells; these compounds are then consumed by the nudribranch and taken up into their dermis, thus generating the nudibranchs' multitude of colorations (Burgess and Eagleston, 2016a, 2017). This also occurs with the sea pens' nematocysts, thus providing the nudibranchs with further defenses. Nudibranchs prefer anywhere from the intertidal region to the sea floor 80 m deep (Burgess and Eagleston, 2019).

There is also a miniature nudibranch, the British Columbian doto (*Doto columbiana*); the adults are just 8 mm in length. As with land snails and slugs, they have sensory antennae on their heads that can detect scents in the surrounding water to find food; these can be quickly withdrawn if a predator attacks (Burgess and Eagleston, 2016b).

Other nudibranchs of the Pacific Northwest include the clown nudibranch (*Triopha catalinae*) a dorid, Taylor's sea hare (*Phyllaplysia taylori*), and the sea lemon (*Peltodoris nobilis*).

Abalone

The abalone is a relative giant among the gastropod mollusks and is classified a basal marine snail of the genus *Haliotis*; some species may live for at least 70 years (Haaker et al., 2001). In the eastern Pacific, they range in adult size from between 15 to 30 cm (Hoiberg, 1993). Abalone are characterized by having between three to seven open holes in their shells (respiratory pores) and a distinctive lining of mother-of-pearl (nacre) in their shells, thus contributing to their commercial and decorative desirability (Haaker et al., 2001). There are seven species found off the coast of western North America, each occupying a different habitat. The only abalones that tolerate the colder waters off the Pacific Northwest are the flat abalone (*H. walallensis*) and the pinto abalone (*H. kamtschatkana kamtschatkana*); the latter is the only abalone found in Puget Sound and the Salish Sea. Populations of pinto abalone have declined by 97% since 1992 and is considered to be a Species of Concern by the US federal government but to be Endangered by the IUCN (Rogers-Bennett, 2007; WDFW, 2019; McDougall et al., 2005). There is currently a pilot program to release hatchery-bred pinto abalone and to track its success rate (Carson et al., 2019). Harvesting of abalone has always been undertaken by divers, never by net, but overfishing and disease has taken many species, such as the white abalone, to near-extinction by the mid-20th century (Haaker et al., 2001). Commercial fishing of the pinto abalone was never permitted in the state of Washington and recreational fishing was outlawed in 1994 (Hester et al., 2011).

Cephalopods (Octopi and Squid)

Squid and octopi, despite their appearance, are also mollusks. They can change colors quickly. While octopi were common in the Pacific Northwest in the 19th century, today they are rare. One species of octopus live in the Sound. With an arm span of over 60–90 cm is the giant Pacific octopus (*Enteroctopus dofleini*), which can grow to nearly 5 m in length. The Humboldt squid (*Dosidicus gigas*) periodically arrives by the thousands in the Strait of Juan de Fuca and the Salish Sea. They weigh about 7–27 kg. Most squid caught for eating are California market squid or opalescent inshore squid (*Doryteuthis [Loligo] opalescens*). They are occasionally seen near the docks, particulalrly at night, when the dock lights are on (Jamison, ND).

CRUSTACEANS

Crustaceans are a very large group of arthropods that includes crabs, lobsters, crayfish, shrimp, krill, woodlice, and barnacles. About 67,000 species are known. As with the other arthropods, they all have exoskeletons that must be shed for the animal to grow.

Crabs and Lobsters

Crabs are decapod crustaceans with a thick exoskeleton and a pair of pincers. Crabs live in both salt and fresh water around the world. Multiple species live in the Puget Sound region. The Dungeness crab (*Cancer magister*) is the best-known crab in the Puget Sound region. It is a favorite dish for residents and tourists alike. The native Puget Sound king crab (*Lopholithodes mandtii*) can be up to 25 cm wide and prefers to live in deep water. A close relative is the Puget Sound box crab (*Lopholithodes foarminatus*), also a deepwater resident, has an opening (foramen) formed by matching semicircular notches in the claws and first walking legs—when the legs are folded tightly, water enters the gill cavity through this opening (WDFW, 2021c).

The signal crayfish (*Pacifastacus leniusculus*) is a freshwater lobster and is larger than the Louisianna crayfish (*Procambarus clarkii*); it is not harvested on a commercial scale, but is allowed to be taken if it has a minimum legal size of 3.5 inches, when it is at least three years old and has already been able to reproduce. It can grow to 15 cm or more in five to six years (WDFW, 2021d). The spiny crayfish (*Euastacus* spp.) are invasive species (WDFW, 2021e).

Shrimp

Shrimp are decapod crustaceans with elongate abdomens. In this way, they are more similar to lobsters than crabs. Their many swimmerets allow them to swim well. There are thousands of species, and they are an important food crop.

The Sound contains multiple species. Sand shrimp (also called grass shrimp) (*Crangon* spp.) are semi-transparent with black spots. They eat smaller shrimp, amphipods, clams, and plants. This is a native species. They are caught now mostly for bait. Sand shrimp are sensitive to salinity and temperature. They like brackish water with a salt concentration of 14–24 parts per thousand and about 18°C. They will migrate for miles to find this combination. They are protandrous hermaphrodites. That is, they begin life as males for a year and then change to females.

Snapping shrimp (family Alpheidae) create an audible sound that can be heard in the bay. In fact, it is second only to the echolocation clicks of sperm whales in volume. They use specialized claws to make the sounds, stun prey, and to communicate. The invasive sand ghost shrimp (*Biffarius arenosus*) is pale in color and grows to about 13 cm. It is a filter feeder that lives in a burrow and reworks the bay bottom somewhat like earthworms do for soil. They disturb oyster beds.

ECHINODERMS

Sea Stars

Sea stars (also known as starfish) are echinoderms of the superphylum Deuterostomia. About 1500 species are known, and they are found around the world. They feature a central disc and usually five arms, but some species have many arms. Their colors vary. Although they seem peaceful, they are actually predators. The pink short-spined sea star (*Pisaster brevispinus*) is pink. It eats mussels, other bivalves, and sand dollars. It is found in the Rosario Strait and Guemes Channel since it needs a high level of salinity. The brittle star has five thin very flexible arms. It is brown or gray and eats detritus and plankton.

Since 2013, sea stars have been suffering from a disease called sea star wasting disease (Jaffe et al., 2019). More than 20 species have been affected. The disease causes necrotic lesions, twisted rays, ray loss, and death. It seems to be viral, but the causative agent has not been identified.

Jaffe et al. (2019) examined the disease in *Leptasterias* spp. This small sea star broods its young rather than releasing them as planktonic larvae, and thus, they might be more susceptible to a viral disease. In 2010, this sea star was common in the Pacific Northwest. The researchers found that this sea star suffered symptoms similar to larger sea stars (Bates et al. 2009). The loss of this species is not a good sign for the general health of Puget Sound. Twenty other species were also in decline.

Harvell et al. (2019) studied the loss of the common predatory sunflower sea star (*Pycnopodia helianthoides*) throughout its traditional range along the West Coast. Sea star wasting disease is responsible, and they found that its outbreaks were associated with unusually warm surface temperatures.

The Puget dwarf brittle star (*Amphipholis pugetana*) is one of the smallest of the brittle stars; they differ from the sea stars in that their arms are very thin and the body mass is smaller in relation to its size. The tube feed are mainly used for digging and burrowing in the sand (Jamison, ND).

PLANTS

INTRODUCTION

As noted earlier, the Puget Sound region comprises areas of Mediterranean-type ecosystems, temperate rain forests, montane environments, as well as marine areas, for many plants and fungi (Kottek et al., 2006). With this variety of habitats, the Puget Sound region has a diverse array of plants. Here we will offer a representative sampling of those. For a more extensive listing of the plants of the region, we recommend the extensive catalog of vascular plants prepared by the Burke Museum at the University of Washington (Burke Museum, 2021). The ornamental species planted by landscapers, businesses, and homeowners are too numerous to mention here, given that our aim was to describe the native environment. The reader should refer to other works that give more weight to such a topic.

Plants are classified into a number of groups. The most primitive are the mosses, liverworts, and hornworts, more advanced plants include ferns and their allies (such as glubmosses and horsetails), the most complex plants are the vascular forms, such as tree ferns, gymnosperms, and angiosperms (flowering plants) that include both dicotyledons and monocotyledons (Whittaker, 1969; Margulis, 1971).

Plants Associated with Water

Puget Sound has long been surrounded by marshland, wetland, and tidal flats that were regularly inundated by tidal action. However, much of the low-lying shoreline at the edges of the Sound has since been developed for housing and commercial use. This has resulted in a transition from the native pickleweed (*Salicornia depressa*) to more competitive cordgrass (*Spartina* spp.), the more salt-tolerant native salt grass (*Distichlis spicata*), and Silvery saltbush (*Atriplex argentea*) (EOPS, 2021b).

Eelgrass (*Zostera marina*) provides a home to Taylor's sea hare (*Phyllaplysia taylori*), a type of nudibranch, and a spawning habitat for the Pacific herring (*Clupea haengus pallasi*) and for outmigrating juvenile salmon (*Oncorhyncus* spp.) (Ort et al., 2012). Below the surface, at an average depth of about 4 m, the soil is rich in carbon, providing a large storage system of carbon in Puget Sound (see Figure 8.2) (Collins, 2020).

Native Land Plants

Native plants, such as the golden paintbrush (*Castilleia levisecta*) or phantom orchid (*Cephalanthera austiniae*) are listed as critically endangered species, mainly due to land-use change, new developments and off trail/road walking and vehicle (e.g., motorcycles, mountain bikes) habitat degradation (Washington Native Plant Society, 2018).

Gymnosperms

The evergreen conifers (gymnosperms) lack flowers and they produce their seeds in cones. They evolved about 300 million years ago. The most familiar trees of Washington State are the conifers represented by cypresses (Cupressaceae), pines (Pinaceae), and yew trees (Taxaceae).

FIGURE 8.2   Eelgrass in near Bainbridge Island. This area is a key habitat for juvenile forage fish, such as surf smelt (*Hypomesus pretiosus*). They are food for salmon, orca, and marine birds (Photograph courtesy of David Ayers and US Geologic Survey).

Cupressaceae

The western red cedar (*Thuja plicata*) is of particular importance to the Native Americans and First Nations of the Pacific Northwest, as a source of building material for their long houses, totem poles, and dugout canoes.

Junipers are also found throughout the Puget Sound region, exemplified by the western juniper (*Juniperus occidentalis*) and common or mountain juniper (*J. communis*).

Pinaceae

Pines vary in size from shrubs to very tall trees. They evolved about 200 million years ago. They are native to the northern hemisphere with only a few examples in the tropics of the southern hemisphere. Pines hybridize easily, and this characteristic complicates their evolutionary history.

Two very common pines of the slopes of the ranges surrounding Puget Sound are the Sitka spruce (*Picea sitchensis*) and the Pacific silver fir (*Abies amabilis*); other memorable types of pines include the western white pine (*Pinus monticola*), lodgepole pine (*Pinus contorta*), western hemlock (*Tsuga heterophylla*), Douglas fir (*Pseudotsuga menziesii*), western larch (*Larix occidentalis*), and noble fir (*Abies procera*). The noble fir is considered by many to be the perfect tree to decorate around the winter holiday and New Year celebrations.

Taxaceae

There are two species of yew found in the Pacific Northwest: the Pacific or western yew (*Taxus brevifola*) and the non-native English yew (*T. baccata*), but the latter is predominantly found only as an ornamental plant. Traditional medicine the world over has used the bark of the yew for medicaments, including as an anti-inflammatory, an analgesic, anticonvulsant, antipyretic, antifungal, antibacterial, and more recently as the source of the anticancer drug taxol (Juyal et al., 2014).

The eastern slopes of the Olympic Mountains host predominantly western hemlock with some Pacific silver fir (Gavin et al., 2013). Most of the Cascades's lower and middle elevations are covered in coniferous forest; the higher altitudes have extensive meadows as well as alpine tundra and glaciers. The western slopes of the North Cascades are densely covered with Douglas-fir (*Pseudotsuga menziesii*), western hemlock (*Tsuga heterophylla*), and red alder (*Alnus rubra*) (Mueller and Mueller, 2002). Forests of large, coniferous trees (western red cedars, Douglas-firs, western hemlocks, firs, pines, spruces, dominate most of the Cascade Range.

ANGIOSPERMS

Angiosperms, which include flowering plants, grasses and sedges are the most diverse group of terrestrial plants. Compared with the evergreen gymnosperms, they are deciduous, meaning that they shed their leaves with the onset of autumn. Their ancestors diverged from gymnosperms about 300 mya (early Permian Period), as determined by DNA evidence (De La Torre et al., 2017). They are characterized by having flowers, endosperm within their seeds, as well as production of fruit bearing the seeds (Raven et al., 2005).

The core angiosperms can be placed into at least three groups: magnoliids, monocots, and eudicots. Mangnoliids and eudicots differ from monocots in that newly spouted seeds emerge with a pair of cotyledons (embryonic leaves that emerge first to capture sunlight) whereas the monocots have only a single cotyledon.

Oaks are important trees in the Pacific Northwest. Their acorns were a large source of carbohydrate and protein for Native American tribes. The Oregon white oak (*Quercus garryana*) is found west of the Cascades, throughout the Puget Sound lowlands, northeastern Olympic Peninsula, and the islands of the Salish Sea (Franklin and Dyrness, 1988).

Some trees tend to live in habitats near rivers, such as the white alder (*Alnus rhombifolia*), and black cottonwood (*Populus trichocarpa*). Another plant that proliferates during the spring and early summer in newly watered environments is the skunk cabbage (*Lysichiton americanus*); their leaves are the largest of any native plant in the region, 30–150 cm long and 10–70 cm wide when mature (Giblin, 2015).

Other trees and tree-like shrubs include the red alder (*Alnus rubra*), Pacific madrone (*Arbutus menziesii*), bigleaf maple (*Acer macrophyllum*), and western blue elderberry (*Sambucus nigra cerulea*). There are two pockets of native quaking aspen (*Populus tremuloides*) found in the Puget Sound region, one on the southeastern tip of Vancouver Island, the other at the southern end of the Hood Canal.

Shrubs

In addition to trees, a number of shrubs are also found in the Puget Sound region. For example, and sagebrush (*Artemesia tridentata*) and manzanita (*Arctostaphylos* spp.) are amongst the species of shrubs and small trees in the prairie regions of Puget Sound. The word manzanita means "little apple" in Spanish. They do well in poor soil and need little water. The bark is red or orange, and the branches are twisted. The berries and flowers of most are edible.

Another native shrub is the vine maple (*Acer circinatum*) and is closely related to the Japanese fullmoon maple (*A. japonicum*) and the Korean maple (*A. psuedosieboldianum*) being the only member of its genus known to live outside Asia.

Coyote bush (*Baccharis pilularis*) is a very common plant throughout much of Washington State. It is a secondary pioneer plant in communities, such as coastal sage scrub and chaparral. It does not do well in shade and is easily replaced by other species that grow taller. One species of purple sage (*Salvia dorii*) is native to the Puget Sound region. In addition, hybridization is common. Some sage plants have trichomes (hairs) on the leaves that release an oil with a scent that makes them undesirable to animals and insects. Pacific poison oak (*Toxicodendron diversilobum*) and sumac (*Rhus glabra*) contain an oil called urushiol that causes an allergic reaction in many people. Both poison oak and sumac are widespread throughout the region. Other shrubs include service-berry (*Amelanchier utahensis* and *A. alnifolia*), creeping snowberry (*Symphoricarpos mollis*), honeysuckle (*Lonicera* spp.), huckleberry and blueberry (*Vaccinium* spp.), and nettle (*Urtica* spp.).

Flowering Plants

Many flowering plants are found in the Puget Sound region. For example, the state flower, the Pacific rhodedendron (*Rhododendron macrophyllum*), is a common plant with beautiful pink flowers. Narrowleaf milkweed (*Asclepias fascicularis*) is a perennial with lavender or whitish flowers. The seedpods split open to release seeds with silky hairs. These plants are important larval host plants for Monarch butterflies. Other flowering plants include the beardtongue (*Penstemon* spp.), and mountain bluecurls (*Trichostema oblongum*). However, the list of flowering plants is enormous, and even the desert areas feature many flowers, especially in the spring.

Meadow Plants

Many of the small plants found in meadows belong to the legume family, Fabaceae. Examples include the American bird's-foot trefoil (*Acmispon americanus*), milk-vetch (*Astragalus* spp.), lupin (*Lupinus* spp.), and clover

(*Trifolium*spp.) (Burke Museum, 2021). Legumes are noted for having root nodules that comprise symbionic and mutualistic nitrogen-fixing bacteria, rhizobia, thus enabling the plants to grow in soils under nitrogen-poor conditions (Zahran, 1999). Recent evidence indicates that all the legumes are monophyletic, in other words, they all derive from a single common ancestor (Lewis et al., 2005).

Grasses and Sedges

Grasses and sedges (monocotyledons) are the most important food crop in the world. In addition, they are used for forage and decorative plants. They are found in many habitats and are the most common plant in the world. The Puget Sound region features multiple types of native grasses, including purple three-awn (*Aristida purpurea*), alpine fescue (*Festuca brachyphylla*), Idaho fescue (*Festuca idahoensis*), red fescue (*Festuca rubra*), prairie Junegrass (*Koeleria macrantha*), Vancouver wildrye (*Leymus vancouverensis*, a sterile hybrid of *L. molis* and *L. triticoides*), bearded melic grass (*Melica aristata*), Indian white-grain mountain-rice grass (*Oryzopsis asperifolia*), and pine bluegrass (*Poa secunda*). A number of grass-like plants are also important. They include the sedges (*Carex* spp.), rushes (*Juncus* spp.), shore blue-eyed grass (*Sisyrinchium littorale*), and golden blue-eyed grass (*Sisyrinchium californicum*).

FERNS

The ferns are vascular plants that reproduce by using spores. They lack flowers and seeds. Their leaves are complex, and most produce the fiddleheads that are widely recognized. Ferns occupy a wide range of habitats, but they tend to occur in those areas in which flowering plants are limited, such as shady areas, rock faces, and acid wetlands. Some are considered weeds, and others grow as epiphytes. Common examples in the Puget Sound region include the polypody ferns (*Polypodium* spp.), native western sword ferns (*Polystichum munitum)*, giant chainfern (*Woodwardia fimbriata*), goldback ferns (*Pteridium* spp.), wood ferns (*Dryopteris arguta*), and maidenhair ferns (*Adiantum aleuticum*).

NON-VASCULAR PLANTS

The Puget Sound region also includes quite a number of non-vascular plants. They are somewhat easy to overlook since the vascular plants now dominate most habitats. However, the non-vascular plants can still be found in various areas, including road cuts, rock outcrops, decaying logs, stream banks, and more. Others live as epiphytes in trees. The nonvascular plants form three divisions: mosses (Bryophyta), liverworts (Hepatophyta), and hornworts (Anthocerotophyta). They all lack vascular tissue and organs (e.g., stems, roots, leaves, flowers). Some have leaf-like structures, but they are not true leaves. In place of roots, they have rhizoids that lightly anchor them into the soil. They include generalized cells called parenchyma, growth occurs in the meristem, and they are covered by a cuticle. The gametophyte is the dominant stage. The mosses (Bryophyta) in the Puget Sound region include *Tortula mucronifolia, Timmia austriaca* and *Scleropodium obtusifolium.* The liverworts (Hepatophyata) are represented by *Cephaloziella* spp., *Lophocolea heterophylla,* and *Porella cordeana.* An example of a hornwort (Anthrocerotae) in the Puget Sound region is *Anthoceras punctatus* (Calabria et al., 2015).

FUNGI

The Puget Sound region has a wide array of fungi. They belong to two major groups. The Basidiomycetes include common edible mushrooms. The fleshy structure that we eat is the fruiting body that is called the basidium. The main body of the fungi exists underground as a system of mycelium. The basiciomycetes include the common *Agaricus* of pizza fame, boletes, puffballs, and polypores. The other main group is the Ascomycetes, which form a structure called an ascus. These include morels, truffles, and cup fungi.

Common fungi of the Puget Sound region include *Stropharia ambigua,* deadly skullcap (*Galerina marginata*), russet toughshank (*Gymnopus dryophilus*), beaked earthstar (*Geastrum pectinatum*), the Prince (*Agaricus augustus*), the fairy ring mushroom (*Marasmius oreades*), and the death cap (*Amanita phalloides,* very poisonous).

ALGAE

The Puget Sound kelp forests feature an amazing array of organisms. One of the larger organisms is the giant kelp (*Macrocystis pyrifera*), the largest marine macroalga in the world whose photosynthesis provides nourishment to a large ecosystem. The kelp forest comprises stripes (stalks) and fronds (leaves) up to 200 m in height and which provide shelter to hundreds of species (Foster and Schiel, 1985). Under optimal conditions, they may grow at a rate of 30–60 cm per day (National Ocean Service, 2013, 2020).

INVASIVE SPECIES

Invasive species can have a detrimental effect on native species. The movement of humans from continent to continent provides an excellent opportunity for plants and animals to move from one region to another. For example, the shothole borer (*Scolytus rugulosus*) is a boring beetle that drills into tree trunks and branches, bringing with it a pathogenic fungus along with other fungi that are conducive to establishing and nurturing shothole borer colonies. Shothole borers are known to have attacked more than 200 species of native, exotic, and agricultural trees in the Pacific Northwest and have been found in a number of environments—from urban landscapes to commercial groves (Doerr and VanBuskirk, 1993).

CLIMATE CHANGE

Climate change has been having an effect upon both native and non-native species; changes such as earlier spring growth of Douglas-fir, reduction in habitat for the northern spotted owl, increased prairie expansion in the Puget Sound, and montane meadows may decrease as reduced snowpack and warmer weather enable trees to proliferate and become established; the extent of effects upon garry oak habitat is unpredictable (Mauger et al., 2015).

Some extraordinary phenomena have recently occurred in the Pacific Northwest. During June of 2021, an unusually strong and persistent ridge of high pressure, a heat dome, enveloped the northwestern United States and Canadian British Columbia. Record daily temperatures ensued, at times exceeding 42°C in Seattle, Washington, and 48°C in Lytton, British Columbia, record highs since records began in 1940 (NWS, 2021). There is predicted to be a concomitant period of increased evaporation of lakes and waterways, resulting in loss of habitat for many animals and plants, as well as loss of water volume in the streams for salmon and trout spawning.

REFERENCES

Abadía-Cardoso A, Freimer NB, Deiner K, Garza JC (2017) Molecular population genetics of the Northern Elephant Seal *Mirounga angustirostris. Journal of Heredity* 108(6): 618–627.

ADFG (ND) *American Mink Species Profile, Alaska Department of Fish and Game, Juneau, Alaska.* Retrieved from: https://www.adfg.alaska.gov/index.cfm?adfg=americanmink.main. June 22, 2021.

Agrawal A (2017) *Monarchs and Milkweed: A Migrating Butterfly, a Poisonous Plant, and Their Remarkable Story of Coevolution.* Princeton University Press, Princeton, New Jersey.

Akerman JT, Peterson SH (2017) California gull diet, movements, and use of landfills in San Francisco Bay. *Tideline* 40: 1–2.

Anderson EM, Lovallo MJ (2003) Bobcat and lynx. In: Feldhamer GA, Thompson BC, Chapman JA (editors). *Wild Mammals of North America: Biology, Management, and Conservation* (2nd ed.), pp. 758–786. Johns Hopkins University Press, Baltimore, MD.

Arp AJ, Hansen BM, Julian D (1992) Burrow environment and coelomic fluid characteristics of the echiuran worm *Urechis caupo* from populations at three sites in northern California. *Marine Biology* 113: 613–623.

Aubry KB, Statham MJ, Sacks BN, Perrine JD, Wisely SM (2009) Phylogeography of the North American red fox: Vicariance in Pleistocene forest refugia. *Molecular Ecology* 18: 2668–2686.

Audubon (ND) *Guide to North American Birds.* Bald Eagle. Retrieved from: www.audubon.org/field-guide/bird/bald-eagle. March 8, 2021.

Azanza B, Rossner G, Ortiz-Jaureguizar E (2013) The early Turolian (late Miocene) Cervidae (Artiodactyla, Mammalia) from the fossil site of Dron-Durkheim 1 (German) and implications on the origin of crown cervids. *Paleobiodiversity and Paleoenvironments* 93(1): 217–258.

Baker GM (1999) *Naturalised terrestrial Stylommatopha (Mullusca: Gastropoda). Fauna of New Zealand, Number 38.* Manaaki Whenua Press, Canterbury, New Zealand. p. 254.

Balcomb KC, Bigg MA (1986) Population biology of three resident killer whale pods in Puget Sound and off southern Vancouver Island. In: Kirkevold BC, Lockard JS (editors). *Behavioral Biology of Killer Whales*, pp. 85–95. Alan R. Liss, Inc., New York, NY.

Barrat J (2013) *Suburban Raccoons More Social Yet Dominance Behavior Remains That of a Solitary Animal*. Smithsonian Insider. Retrieved from: https://insider.si.edu/2013/07/suburban-life-does-not-alter-solitary-ways-of-the-raccoon/. September 27, 2020.

Bates AE, Hilton BJ, Harley CDG (2009) Effects of temperature, season and locality on wasting disease in the keystone predatory sea star Pisaster ochraceus. *Diseases of Aquatic Organisms* 86:245–251.

BBWMA (2021) Boundary Bay Wildlife Management Area. Retrieved from: https://www2.gov.bc.ca/gov/content/environment/pklants-animals-ecosystems/wilflife/wildlife-habitats/conservation-lands/wma/wmas-list/boundary-bay. June 28, 2021.

Beamer EM, McBride A, Greene C, Henderson R, Hood G, Wolf K, Larson K, Rice C, Fresh K (2005) *Delta and Nearshore Restoration for the Recovery of Wild Skagit River Chinook Salmon: Linking Estuary Restoration to Wild Chinook Salmon Populations*. Skagit River System Cooperative, LaConner, Washington.

Behnke R (2010) *Trout and Salmon of North America*. Simon and Schuster, New York, NY.

Behnke RJ, Tomelleri JR (illustrator) (2002) *Genus Oncorhynchus*. Trout and Salmon of North America. The Free Press, Apollo Beach, Florida.

Bergamin A (2014) *Nudibranchs, Kings of the Tidepool, Command an Audience*. Bay Nature. Retrieved from: https://baynature.org/article/nudibranchs-kings-tidepool/.

Beschta RL, Ripple RJ (2009) Large predators and trophic cascades in terrestrial eco-systems of the western United States. *Biological Conservation* 142: 2401–2414.

BirdLife International (2016) *Rallus Obsoletus*. IUCN Red List of Threatened Species. September 15, 2020.

BMNHC (2006) *Bushy-Tailed Woodrat, Neotoma cinerea*. Burke Museum of Natural History and Culture, University of Washington, Seattle, Washington.

Brand CJ, Keith LB (1979) Lynx demography during a snowshoe hare decline in Alberta. *The Journal of Wildlife Management* 43(4): 827–849.

Brown CW, Stebbins RC (1964) Evidence for hybridization between the blotched and unblotched subspecies of the salamander *Ensatina eschscholtzii*. *Evolution* 18: 706–707.

Buhler B (2018) *There Are so Many Scorpions*. Bay Nature, Winter 2018. Retrieved from: https://baynature.org/article/there-are-so-many-scorpions/. accessed July2021.

Burgess D, Eagleston A (2016a) *The Slender Sea Pen. Eyes Under Puget Sound – Critter of the Month*. Department of Ecology, State of Washington, Tacoma, Washington. Retrieved from: https://ecology.wa.gov/Blog/Posts/March-2016/Eyes-under-Puget-Sound-Critter-of-the-Month--Slender-sea-pen. July 9, 2021.

Burgess D, Eagleston A (2016b) *Sea Slugs: The British Columbian Doto. Eyes Under Puget Sound – Critter of the Month*. Department of Ecology, State of Washington, Tacoma, Washington. Retrieved from: https://ecology.wa.gov/Blog/Posts/March-2016/Eyes-under-Puget-Sound-Critter-of-the-Month. July 10, 2021.

Burgess D, Eagleston A (2017) *Get Ready to "Fall" for the Orange Sea Pen. Eyes Under Puget Sound – Critter of the Month*. Department of Ecology, State of Washington, Tacoma, Washington. Retrieved from: https://ecology.wa.gov/Blog/Posts/September-2017/Eyes-under-Puget-Sound-Critter-of-the-Month--Orange-sea-pen. July 9, 2021.

Burgess D, Eagleston A (2019) *The Striped Nudibranch: Don't Mess with This Ferocious Sea Slug! Eyes Under Puget Sound – Critter of the Month*. Department of Ecology,

State of Washington, Tacoma, Washington. Retrieved from: https://ecology.wa.gov/ Blog/Posts/July-2019/The-striped-nudibranch--Don't-mess-with-this-ferocious-sea-slug!. July 9, 2021.

Burke TE (2013) *Land Snails and Slugs of the Pacific Northwest*. Pacific Northwest Shell Club, Oregon State University Press,Corvalis, Oregon.

Burke Museum (2021a) Mammals of Washington, Burke Museum, University of Washington Herbarium, University of Washington, Seatlle, Washington. Retrieved from https:// www.burkemuseum.org/collections-and-research/biology/mammalogy/mamwash/ rodentia.php, 17 October 2021.

Burke Museum (2021b) *Washington Flora Checklist*. Burke Museum, University of Washington Herbarium, University of Washington, Seatlle, Washington. Retrieved from: https://biology.burke.washington.edu/herbarium/waflora/checklist.php. July 11, 2021.

Calabria LM, Arnold A, Charatz E, Eide G, Hynson LM, Jackmond G, Nannes J, Stone D, Vellella J (2015) A checklist of soil-dwelling bryophytes and lichens of the South Puget Sound prairies of western Washington. *Evansia* 32(1): 30–41.

California Herps (2021) *A Guide to the Amphibians and Reptiles of California, San Diego Gophersnake*, Retrieved from: www.californiaherps.com/snakes/pages/p.c.annectens. html. March 9, 2021.

Cannon Y (2006) Early storage and sedentism on the Pacific Northwest coast: Ancient DNA analysis of salmon remains from Namu, British Columbia. *American Antiquity* 71(1): 123–140.

Carlisle JD, Skagen SK, Kus BE, Riper CV, Paxtons KL, Kelley JF (2009) Landbird migration in the American west: Recent progress and future research directions. *Condor Ornithological Applications* 111: 211–225. DOI: 10.1525/cond.2009.080096.

Carlton JT (1979) *History, Biogeography, and Ecology of the Introduced Marine and Estuarine Invertebrates of the Pacific Coast of North America*. Ph.D. thesis, University of California, Davis, CA. pp. 253–266.

Carson HS, Morin DJ, Bouma JV, Ulrich M, Sizemore R (2019) The survival of hatchery-origin pinto abalone Haliotis kamtschatkana released into Washington waters. *Aquatic Conservation: Marine and Freshwater Ecosystems* 2019: 1–18.

CBD (2016) *The Puget Sound Basin: A biodiversity Assessment*. Center for Biological Diversity and Friends of the San Juans, Portland, Oregon and Friday Harbor, Washington. Retrieved from: www.sanjuans.org/wp-content/uploads/2016/11/ PugetSoundBasinBiodiversityAssessment.pdf. July 7, 2021.

CCDR (2008) *Canada Communicable Disease Report (CCDR) – Vol.34 CCDR-01 – Public Health Agency of Canada*. Phac-aspc.gc.ca. June 13, 2021.

CDC (2012) *Hantavirus*. Centers for Disease Control and Prevention, US Department of Health and Human Services, Atlanta, GA. Retrieved from: https://www.cdc.gov/ hantavirus. June 13, 2021.

CDC (2021) *Western Blacklegged Tick (Ixodes pacificus)*. Center for Disease Control and Prevention, Atlanta, GA. Retrieved from: https://www.cdc.gov/ticks/geographic_ distribution.html. July 10, 2021.

CDFW (2005) *Abalone Recovery and Management Plan*. California Fish and Game Commission, California Department of Fish and Game, Sacramento, CA.

CDFW (ND) *Keep Me Wild: Bobcat*. California Department of Fish and Wildlife. Retrieved from: https://wildlife.ca.gov/Keep-Me-Wild/Bobcat.

Chen K, Baxter T, Muir WM, Groenen MA, Schook LB (2007) Genetic resources, genome mapping and evolutionary genomics of the pig (*Sus scrofa*). *International Journal of Biological Science* 3(3): 153–165.

Chinery M (1993), *Insects of Britain & Northern Europe* (3rd ed.). HarperCollins, London, UK.

Christian CE (2001) Consequences of a biological invasion reveal the importance of mutualism for plant communities. *Nature* 413: 635–639.

Cifelli RL (2000) Cretaceous mammals of Asia and North America, Paleont. *Soc. Korea Special Publication* 4: 49–84.

Cohen AN (2011) *Clathria Prolifera* (Ellis & Solander, 1786). *The Exotics Guide: Non-native Marine Species of the North American Pacific Coast.* Center for Research on Aquatic Bioinvasions. Richmond, California and San Francisco Estuary Insitute, Oakland, California. July 10, 2021.

Collins D (2020) *Sponges of Puget Sound.* Made in Puget Sound, Seattle, Washington. Retrieved from: https://www.madeinpugetsound.org/blog/sponges-of-puget-sound. July 10, 2021.

Conroy CJ, Rowe KC, Rowe KMC, Kamath PL, Aplin KP, Hui L, James DK, Moritz C, Patton JL (2012) Cryptic genetic diversity in *Rattus* of the San Francisco Bay region, California. *Biological Invasions* 15: 741–758.

Corn PS, Bury RB (1991). *Terrestrial Amphibian Communities in the Oregon Coast Range.* In: Ruggiero K, Aubry B, Carey AB, Huff MH (editors). *Wildlife and Vegetation of Unmanaged Douglas-fir Forests,* pp. 304–317. USDA Forest Service, Pacific Northwest Research Station, Olympia, Washington.

COSEWIC (2002) COSEWIC *Assessment and Update Status Report on the Grizzly Bear Ursus arctos in Canada.* Committee on the Status of Endangered Wildlife in Canada, Ottawa, Ontario, Canada. p. 91.

COSEWIC (2010) COSEWIC *Assessment and Status Report on the Western Brook Lamprey Lampetra richardsoni Morrison Creek Population in Canada.* Species at Risk Public Registry, Committee on the Status of Endangered Wildlife in Canada (COSEWIC), Ottawa, Ontario, Canada.

Cox BA (1987) *Native People, Native Lands: Canadian Indians, Inuit and Métis.* McGill-Queen's Press, Montreal, Quebec, Canada. p. 174.

CRITFC (2021) *Tribal Salmon Culture, Salmon Culture if the Pacific Northwest Tribes.* Columbia River Inter-Tribal Fish Commission, Portland, Oregon. Retrieved from: https://www.critfc.org/salmon-culture/tribal-salmon-culture. June 29, 2021.

Cronin MA, Armstrup SC, Garner ER, Vyse GW, Vyse ER (1991) Interspecific and intraspecific mitochondrial DNA variation in North American bears (Ursus). *Canadian Journal of Zoology* 69: 2985–2992.

De La Torre AR, Li Z, Van de Peer Y, Ingvarsson PK (June 2017). Contrasting rates of molecular evolution and patterns of selection among gymnosperms and flowering plants. *Molecular Biology and Evolution* 34(6): 1363–1377.

Dobson AP, Rodriguez JP, Roberts WM, Wilcove DS (1997) Geographic distribution of endangered species in the United States. *Science* 275: 550–553.

Doerr M, VanBuskirk P (1993) *Shothole Borers.* WSU Tree Fruit, Washingto State University, Pullman, Washington. Retrieved from: https://www.treefruit.wsu.edu/crop-protection/opm/shothole-borers/. July 12, 2021.

Dragoo JW, Honeycutt RL (1997) Systematics of mustelid-like carnivores. *Journal of Mammalogy* 8: 426–443.

DU (2010) *The American Wigeon.* Ducks Unlimited, May–June 2010. Retrieved from: https://www.ducks.org/hunting/waterfowl-id/american-wigeon. June 24, 2021.

Duggan T (2021) *Bodega Lab a Key Part of Effort to Save Abalone.* San Francisco Chronicle, San Francisco, CA.

Economist (2012) *Commercial Whaling: Good Whale Hunting.* The Economist. March 4, 2012.

Edvenson JC (1994) Predator control and regulated killing: A biodiversity analysis. *UCLA Journal of Environmental Law and Policy* 13: 31–86.

Eiler KC, Banack SA (2004) Variability in the alarm call of golden-mantled ground squirrels (*Spermophilus lateralis* and *S. saturatus*). *Journal of Mammalogy* 85(1): 43–50.

Elliott J, Holmes K, Chambers R, Leon K, Wimberger P (2008) Differences in morphology and habitat use amount the native mussek *Mytilus trossulus*, the non-native *M. galloprovincialis*, and their hybrids in Puget Sound, Washington. *Marine Biology* 156: 39–53.

EOPS (2021a) *Jellyfish*. Puget Sound Science Review, Encyclopedia of Puget Sound, Puget Sound Institute, University of Washington, Tacoma, Washington. Retrieved from: https://www.eopugetsound.org/science-review/4-jellyfish. July 10, 2021.

EOPS (2021b) *Atriplex argentea*. Encyclopedia of Puget Sound, Puget Sound Institute, University of Washington, Tacoma, Washington. Retrieved from: https://www.eopugetsound.org/species/atriplex-argentea. July 10, 2021.

Federal Records (2021) Seattle Climate Graphs, National Weather Service, NOAA, Seattle Climate Graphs (weather.gov), Seattle, Washington.Retrieved from https://www.wrh.noaa.gov/climate/yeardisp.php?wfo=sew&stn=KSEA&submit=Yearly+Charts; October 17, 2021.

Fedriani JM, Fuller TK, Sauvajot RM, York EC (2000) Competition and intraguild predation among three sympatric carnivores. *Oecologia* 125(2): 258–270.

Feldhamer GA, Thompson BC, Chapman JA (2003) *Wild Mammals of North America: Biology, Management, and Conservation*. JHU Press, Baltimore, Maryland. p. 683.

Fergeson-Leese J, Christie DA, Franklin K, Mead D, Burton P (2001) *Raptors of the World*. Houghton Mifflin Harcourt, Boston, Massachussetts.

Fjeldsa J (1973) Distribution and geographic variation of the Horned Grebe *Podiceps auritus* (Linnaeus, 1758). *Ornis Scandinavica* 4(1): 55–86.

Floyd T (2008) *Smithsonian Field Guide to the Birds of North America*. Harper Collins, New York, NY.

Flynn JJ, Finarelli JA, Zehr S, Hsu J, Nedbal MA (2005) Molecular phylogeny of the Carnivora (Mammalia): Assessing the impact of increased sampling on resolving enigmatic relationships. *Systematic Biology* 54: 317–337.

Foster MS, Schiel DR (1985) The ecology of giant kelp forests in California: A community profile. *US Fish and Wildlife Service Report* 85: 1–152.

Francis WR, Sire de Vilar A (2020) *Bioluminescence and Fluorescence of Three Sea Pens in the North-West Mediterranean Sea*. $bioR_xiv$, Cold Spring Harbor Laboratory, Cold Spring Harbor, New York. DOI: 10.1101/2020.12.08.416396. July 10, 2021.

Franklin and Dyrness (1988) *Natural Vegetation of Oregon and Washington*. Oregon State University Press, Corvalis, Oregon.

Frost R (2015) *Dicamptodon Tenebrosus* (Baird and Girard, 1852). *Amphibian Species of the World: An Online Reference. Version 6.0.* American Museum of Natural History, New York, NY. June 29, 2021.

FS (ND) *Forest Service, San Bernardino National Forest*. Retrieved from: www.fs.usda.gov/detail/sbnf/home/?cid=STELPRD3829099. March 7, 2021.

Gavin DG, Fisher DM, Herring EM, White A, Brubaker LB (2013) *Paleoenvironmental Change on the Olympic Peninsula, Washington: Forests and Climate from the Last Glaciation to the Present. Final Report to Olympic National Park*. University of Washington, Seattle, Washington. p. 109.

Geiger DL, Owen B (2012) *Abalone: Worldwide Haliotidae*. Conchbooks, Hackenhiem, Germany.

Giblin D (editor) (2015) *Lysichiton Americanus*. Burke Museum, University of Washington, Seattle, Washington. July 12, 2021.

Gong AJ (2019) *Voracious Flowers of the Tidepool*. Bay Nature. Retrieved from: https://baynature.org/2019/08/13/voracious-flowers-of-the-tidepool/.

Gould SJ (1977) *Ontogeny and Phylogeny*. Belknap Press, Cambridge, MA.

Grenfell WE Jr (1974) *Food Habits of the River Otter in Suisun Marsh, Central California*. California State University, Sacramento. Retrieved from: http://csus-dspace.calstate.edu/bitstream/handle/10211.9/1554/1974-Grenfell.pdf?sequence=1.

Haaker PL, Karpov K, Rogers-Bennett L, Taniguchi I, Friedman CS, Tegner MJ (2001) *Abalone, California's Living Marine Resources: A Status Report*. California Department of Fish and Game Sacramento, California. pp. 89–97.

Hairston NG, Smith FE, Slobodkin LB (1960) Community structure, population control and competition. *American Naturalist* 94: 421–425.

Hallmann CA, Sorg M, Jongejans E, Siepel H, Hofland N, Schwan H, Stenmans W, Müller A, Sumser H, Hörren T, Goulson D, de Kroon H (2017) More than 75 percent decline over 27 years in total flying insect biomass in protected areas. *PLoS ONE* 12(10): e0185809.

Hallock LA, McAllister KR (2009) *Cope's Giant Salamander*. Washington Herp Atlas. Retrieved from: http://www1.dnr.wa.gov/nhp/refdesk/herp/. June 29, 2021.

Hammerson G (2008) *Rana Draytonii. IUCN Red List of Threatened Species*. IUCN. Retrieved from: http://www.fws.gov/sacramento/es/maps/CRF_fCH_FR_maps/crf_fCH_units.htm. doi: 10.2305/IUCN.UK.2008.RLTS.T136113A4240307.en.

Harrison J (2018) *June Hogs (salmon). Oregon Encyclopedia*. Oregon Historical Society, Portland State University, Portland, Oregon.

Harvell CD, Montecino-Latorre D, Caldwell JM, Burt JM, Bosley K, Keller A, Heron SF, Salomon AK, Lee L, Pontier O, Pattengill-Semmens C, Gaydos JK (2019) Disease epidemic and a marine heat wave are associated with the continental-scale collapse of a pivotal predator (Pycnopodia helianthoides). *Science Advances* 5(1): eaau7042.

Hatch MH (1947) The chelifera and isopoda of Washington and adjacent regions. *University of Washington Publications in Biology* 10(5): 155–274.

Hertzog LA (1990) *Where North Meets South: Cities, Space, and Politics on the United States-Mexico Border*. University of Texas Press, Austin, Texas. pp. 194–201.

Hester JB, Walker JM, Dinnel PA, Schwarck NT (2011) Survey of previously outplanted Pinto (Northern) Abalone (Haliotis kamtschatkana) in the San Juan Island Archipelago, Washington State. In: Pollock NW (editor). *Proceedings of the American Academy of Underwater Sciences 30th Symposium*. AAUS, Dauphin Island, AL.

Hines S (2007) *Waters off Grays Harbor Only Second Place in World Where Glass Sponge Reefs Found*. UW News, University of Washington, Seattle, Washington. Retrieved from: https://www.washington.edu/news/2007/07/30/waters-off-grays-harbor-onlysecond-place-in-world-where-glass-sponge-reefs-found/?menu2=. July 10, 2021.

Hoiberg DH (1993) *Encyclopædia Britannica*. 1: *A-ak Bayes* (15th ed.). Encyclopædia Britannica, Inc., Chicago, Illinois.

Hunter L (2015) *Wild Cats of the World*. Bloomsbury Publishing, London. pp. 146–151.

II (2021) *Washington Insects*. Insect Identification for the Casual Observer. Retrieved from: https://www.insectidentification.org/insects-by-state.php?thisState=Washington. June 30, 2021.

iNaturalist (ND) *Cardiff State Beach California*. Retrieved from: https://www.inaturalist.org/places/cardiff-state-beach#page=3&taxon=47115. February 26, 2021.

IUCN (2015). *Rana Aurora*. IUCN SSC Amphibian Specialist Group, IUCN Red List of Threatened Species, International Union for Conservation of Nature, Cambridge, UK.

Jaffe N, Eberl R, Bucholz J, Cohen CS (2019) Sea star wasting disease demography and etiology in the brooding sea star *Leptasterias* spp. *PLoS ONE* 14(11): e0225248.

Jamison (ND) *Common Puget Sound Plants and Animals*. Retrieved from: https://www.pugetsoundsealife.sseacenter.org/pugetsoundsealife.com/Puget_sound_sea_life/common_plants_and_animals.html. July 10, 2021.

Jeffries SJ, Gearin PJ, Huber, HR, Saul DL, Pruett DA (2000) *Atlas of Seal and Sea Lion Haulout Sites in Washington*. Washington Department of Fish and Wildlife, Wildlife Science Division, Olympia, Washington. p. 150.

Jiang D, Klaus S, Zhang Y-P, Hillis DM, Li J-T (2019) Asymmetric biotic exchange across the Bering land bridge between Eurasia and North America. *National Science Review* 6: 739–745.

Jobling JA (2010) *The Helm Dictionary of Scientific Bird Names*. Christopher Helm, London, UK. pp. 114–128.

Johnson WE, O'Brien SJ (1997) Phylogenetic reconstruction of the Felidae using 16S rRNA and NADH-5 mitochondrial genes. *Journal of Molecular Evolution* 44(suppl 1): S98–S116.

Johnson CS (2013)A New Oyster Invades, Sea Grant California, University of California San Diego, La Jolla, California. Retrieved from https://caseagrant.ucsd.edu/news/a-new-oyster-invades, February 26, 2021.

Jones N (2015) *A Delicate Balance: Protecting Northwest's Glass Sponge Reefs*. Yale School of the Environment, Yale University, New Haven, Connecticut.

Juyal D, Thawani V, Thaledi S, Joshi M (2014) Ethnomedical properties of *Taxus wallachiana* Zucc. (Himalayan Yew). *Journal of Traditional Complementary Medicine* 4(3): 159–161.

Kaser MR, Howard GC (2021) *Making and Unmaking of the San Diego Bay*. CRC Press, Baton Rouge, Florida.

Kemp TS (1987) Fossil synapsids: The ecology and biology of mammal-like reptiles. *Science* 236(4803): 862–863.

Kemp TS (2005). *The Origin and Evolution of Mammals*. Oxford University Press, Oxford, United Kingdom.

Kenyon KW (1969) *The Sea Otter in the Eastern Pacific Ocean*. U.S. Bureau of Sport Fisheries and Wildlife, Washington, D.C.

King JL (2004) *The Current Distribution of the Introduced Fox Squirrel (Sciurus niger) in the Greater Los Angeles Metropolitan Area and its Behavioral Interaction with the Native Western Gray Squirrel (Sciurus griseus)*. Master's thesis, California State University, Los Angeles, California.

Klymkowsky MW, Cooper MM, Begovic E, Lymkowsky R (2016) Sexual dimorphism. *Biology LibreTexts*. September 27, 2020.

Kottek M, Grieser J, Beck C, Rudolf B, Rubel F (2006) World Map of the Köppen-Geiger climate classification updated. *Meteorologische Zeitschrift* 15(3): 259–263. DOI: 10.1127/0941-2948/2006/0130. January 3, 2021.

Krause WJ, Krause WA (2006) *The Opossum: Its Amazing Story* Archived 2012-12-11 at the Wayback Machine. Department of Pathology and Anatomical Sciences, School of Medicine, University of Missouri, Columbia, Missouri. p. 39.

Krebs CJ, Boonstra R, Boutin S, Sinclair ARE (2001) What drives the 10-year cycle of snowshoe hares?: The ten-year cycle of snowshoe hares—one of the most striking features of the boreal forest—is a product of the interaction between predation and food supplies, as large-scale experiments in the Yukon have demonstrated. *BioScience.* 51 (1): 25–35.

Kruckenhauser, L, Pinsker W, Haring E, Arnold W (1999) Marmot phylogeny revisited: Molecular evidence for a diphyletic origin of sociality. *Journal of Zoological*

Systematics and Evolutionary Research 37(1): 49–56. DOI: 10.1046/j.1439-0469.1 999.95100.x.

Kucera T (1997) *California Giant Salamander (Report)*. California Department of Fish and Game, Sacramento, California.

Kutcha SR, Krakauer AH, Sinervo B (2008) Why does the yellow-eyed ensatina have yellow eyes? Batesian mimicry of Pacific newts (Genus *Taricha*) by the Salamander *Ensatina eschscholtzii xanthoptica*. Evolution. DOI: 10.1111/j.1558-5646.2008.00338.x.

Lafferty KD, Tinker MT (2014) Sea Otters are recolonizing southern California in fits and starts. *Ecosphere* 5(5): 1–11. DOI: 10.1890/ES13-00394.1.

Lantz DE (1909) The brown rat in the United States. United States Department of Agriculture Biological Survey, Bulletin 33: 1–54.

Larsen DN (1984). Feeding habits of river otters in coastal southeastern Alaska. *Journal of Wildlife Management* 48: 1446–1452.

Larson G, Dobney K, Albarella U, Fang M, Matisoo-Smith E, Robins J, Lowden S, Finlayson H, Brand T, Willerslev E, Rowley-Conwy P, Andersson L, Cooper A (2005) Worldwide phylogeny of wild boar reveals multiple centers of pig domestication. *Science* 307(5715): 1618–1621.

Leopold A (1949) Thinking like a mountain. In: *A Sand County Almanac: And Sketches Here and There*. Oxford University Press, Oxford, United Kingdom.

Lewis G, Schrire B, Mackinder B, Lock M (2005) (editors) *Legumes of the World*. The Royal Botanic Gardens, Kew, United Kingdom. p. 577.

Liebhold AM, Kean JM (2019) Eradication and containment of non-native forest insects: Successes and failures. *Journal of Pest Science* 92: 83–91.

Lim G (2021) *Seattle Residents Report Coyote Sighting; Wildlife Experts Say it's Not Uncommon*. Q13 Fox Seattle, Seattle, Washington. Retrieved from: https://www.q13fox.com/news/seattle-residents-report-coyote-sightings-wildlife-experts-say-its-not-uncommon. June 16, 2021.

Lister BC, Garcia A (2018) Climate-driven declines in arthropod abundance restructure a rainforest food web. *Proceedings of the National Academy of Sciences USA* 115: E10397–E10406.

Litalien R (2004) *Champlain: The Birth of French America*. Montreal, Quebec, McGill-Queen's Press. pp. 312–314.

Luís C, Bastos-Silveira C, Cothran EG, Oom MdO (2006) Iberian origins of new world horse breeds. *Journal of Heredity* 97(2): 107–113.

Lunsford R, Helmstetler H (2003) *Katharina Tunicata. Phylum Mollusca*. Walla Walla University College Place, Washington.

Lyons AL, Gaines WL, Begley J, Singleton P (2016) *Grizzly Bear Carrying Capacity in the North Cascade Ecosystem*. US Department of Fish and Agriculture, Seattle, Washington.

Macdonald D (2001) Beavers. In: Macdonald D (editor) *The New Encyclopedia of Mammals*, pp. 590–591, 596–597. Oxford University Press, Oxford, United Kingdom.

Maffe WA (2000) A note on invertebrate populations of the San Francisco Estuary. In: Olofson PR (editor). *Baylands Ecosystem Species and Community Profiles: Life Histories and Environmental Requirements of Key Plants, Fish and Wildlife*, pp. 184–192. San Francisco Bay Regional Water Quality Control Board, Oakland, CA.

Mapes LV (2019) *Southern Resident Orcas Spotted in Home Waters off San Juan Island After Unusual Absence*. The Seattle Times, Seattle, Washington. Retrieved from: https://www.seattletimes.com/seattle-news/environment/Southern-resident-orcas-spotted-in-home-waters-off-San-Juan-Island-after-unusual-absence. June 23, 2021.

Margulis L (1971) Whittaker's five kingdoms of organisms: Minor revisions suggested by considerations of the origin of mitosis. *Evolution* 25(1): 242–245.

Markin GP (1969) The seasonal life cycle of the Argentine ant in Southern California. *Annals of the Entomological Society of America* 63(5): 1238.

Marquez P (2013) *A Field Guide to Some Snails & Slugs in Washington for Survey Inspectors.* USDA-APHIS-PPQ, Hawthorne, California.

Martin G (2011) *The Middle Way. Bay Nature.* Retrieved from: https://baynature.org/article/the-middle-way/. September 27, 2020.

Martín-Durán JM, Passamaneck YJ, Martindale Mark Q, Hejnol Andreas (2016) The developmental basis for the recurrent evolution of deuterostomy and protostomy. *Nature Ecology & Evolution* 1: 0005.

Mathiasson ME, Rehan SM (2019) Status changes in the wild bees of Northeastern North America over 125 years Insect Conservation and Diversity 12: 278-288 . DOI: 10.1111/ icad.12347.

Mauger GS, Casola JH, Morgan HA, Strauch RL, Jones B, Curry B, Busch Isaksen TM, Binder LW, Krosby MB, Snover AK (2015) *State of Knowledge: Climate Change in Puget Sound. Report prepared for the Puget Sound Partnership and the National Oceanic and Atmospheric Administration.* Climate Impacts Group, University of Washington, Seattle, Washington.

Mayer JJ (2017) Introduced wild pigs in North America: History, problems and Management. In: Melleti M, Meijaard E (editors). *Ecology, Conservation and Management of Wild Pigs and Peccaries*, pp. 299–312. Cambridge University Press, Cambridge, United Kingdom.

Mayer JJ, Brisbin IL Jr (2008) *Wild Pigs in the United States: Their History, Comparative Morphology, and Current Status.* University of Georgia Press, Georgia. p. 20.

McDougall PT, Ploss J, Tuthill J (2005) *Haliotis Kamtschatkana. 2006 IUCN Red List of Threatened Species.* International Union for Conservation of Nature, Cambridge, United Kingdom.

McKercher L, Gregoire DR (2020) *Lithobates Catesbeianus* (Shaw, 1802). U.S. Geological Survey, Nonindigenous Aquatic Species Database, Gainesville, Florida.

McKey-Fender D, Fender WM, Marshall VG (1994) North American earthworms native to Vancouver Island and the Olympic Peninsula. *Canadian Journal of Zoology* 72(7): 1325–1339.

McPhail JD, Paragamian VL (2000) Burbot biology and life history. In: Paragamian VL, Willis DW (editors), *Burbot: Biology, Ecology, and Management*, pp. 11–23. Fisheries Management Section of the American Fisheries Society, Bethesda, Maryland.

Melquist WE, Dronkert AE (1987) River otter. In: Novak M, Baker JA, Obbard ME, Malloch B (editors). *Wild Furbearer Management and Conservation in North America*, pp. 626–641. Ontario Ministry of Natural Resources, Toronto, Canada.

Mesa M (1994) Effects of multiple acute stressors on the predator avoidance ability and physiology of juvenile chinook salmon. *Transactions of the American Fisheries Society* 123: 786–793.

MFG (ND) Peamouth - *Mylocheilus Caurinus. Montana Field Guide.* Montana Natural Heritage Program and Montana Fish, Wildlife and Parks, Helena, Montana. Retrieved from: https://fieldguide.mt.gov/speciesDetail.aspx?elcode=AFCJB24010. June 30, 2021.

Miller CR, Waits LP (2006) Phylogeography and mitochondrial diversity of extirpated brown bear (*Ursus arctos*) populations in the contiguous United States and Mexico. *Molecular Ecol*ology 15: 4477–4485.

Morris RH, Abbott DP, Haderie EC (1980) *Intertidal Invertebrates of California.* Stanford University Press, Palo Alto, California.

Morse D (2007) *Predator Upon a Flower: Life History and Fitness in a Crab Spider.* Harvard University Press, Cambridge, Massachusetts.

Moyle PB, Israel JA, Purdy SE (2008) *Salmon, Steelhead, and Trout in California: Status of an Emblematic Fauna.* Center for Watershed Sciences, University of California Davis, Davis, California. pp. 89–91.

Mueller M, Mueller T (2002) *Exploring Washington's Wild Areas* (2nd ed.). The Mountaineers Books, Seattle, Washington, p. 99.

Müller F (1878) Ueber die Vortheile der Mimicry bei Schmetterlingen. *Zoologischer Anzeiger* 1: 54–55.

Muller MJ, Storer RW (1999) Pied-billed Grebe (*Podilymbus podiceps*). In: Poole A, Gill F (editors). *The Birds of North America, No. 410.* The Birds of North America, Inc., Philadelphia, Pennsylvania.

Mulvaney D (2013) *Green Atlas: A Multimedia Reference.* SAGE Publications, Thousand Oaks, California. p. 32.

Murray W (2004) *Elsevier's Dictionary of Reptiles.* Elsevier, Amsterdam, Netherlands. p. 122.

National Ocean Service (2013) *What Lives in a Kelp Forest: Kelp Forests Provide Habitat for a Variety of Invertebrates, Fish, Marine Mammals, and Birds NOAA.* September 27, 2020.

National Ocean Service (2020) *Kelp Forests: A Description.* Retrieved from: https://sanctuaries.noaa.gov/visit/ecosystems/kelpdesc.html. September 27, 2020.

Neilson S (2019) *More Bad Buzz for Bees: Record Number of Honeybee Colonies Died Last Winter.* KQED. Retrieved from: https://www.npr.org/sections/thesalt/2019/06/19/733761393/more-bad-buzz-for-bees-record-numbers-of-honey-bee-colonies-died-last-winter#:~:text=Bee%20decline%20has%20many%20causes,systems%20and%20can%20kill%20them.

NOAA (2011) *National Oceanic and Atmospheric Agency: San Diego climate by month.* U.S. Department of Commerce National Oceanic & Atmospheric Administration National Environmental Satellite. Retrieved from: https://www.wrh.noaa.gov/sgx/climate/san-san-month.htm.

NOAA (ND) *Endangered Species Conservation.* NOAA Fisheries. Retrieved from: https://www.fisheries.noaa.gov/topic/endangered-species-conservation#resources. June 29, 2021.

Norris J (ND) *Jellyfish in Puget Sound.* Department of Ecology, State of Washington, Tacoma, Washington. Retrieved from: https://ecology.wa.gov/Research-Data/Monitoring-assessment/Puget-Sound-and-marine-monitoring/Jellyfish. July 10, 2021.

Norwak RM (1999) *Walker's Mammals of the World.* Johns Hopkins University Press, Baltimore, Maryland. p. 1521.

NPS (2018) *Marine Invertebrates.* National Park Service. Retrieved from: https://www.nps.gov/goga/learn/nature/marine-invertebrates.htm.

NPS (ND) *Cougar.* Olympic National Park, National Park Service. Retrieved from: https://www.nps.gov/olym/learn/nature/cougar.htm. June 16, 2021.

NWS (2021) *National Weather Service, Washington DC.* https://www.w2.weather.gov/climate/. July 12, 2021.

Oksanen L, Fretwell SD, Arruda J, Niemala P (1981) Exploitation ecosystems in gradients of primary productivity. *American Naturalist* 118: 240–261.

Ordeñana MA, Crooks KR, Boydton EE, Fisher RN, Lyren LM, Siudyla S, Haas CD, Harris S, Hathaway SA, Tureschak GM, Miles AK, Van Vuren DH (2010) Effects of urbanization on carnivore species distribution and richness. *Journal of Mammalogy* 91(6): 1322–1331.

Ort BS, Cohen S, Boyer KE, Wyllie-Echeverria S (2012) Population structure and genetic diversity among eelgrass (*Zosteria marina*) beds and depths in San Francisco Bay. *Journal of Heredity* 103: 533–546.

Ortiz JL, Muchlinski AE (2014) Urban/suburban habitat use by a native and invasive tree squirrel. *Bulletin of the Southern California Academy of Sciences* 113: 116.

Ostrom JH (1973) The ancestry of birds. *Nature* 242(5393): 136.

Padian K (1986) The origin of birds and the evolution of flight. *Memoirs of the California Academy of Sciences* 8: 1–55.

Parr I (2019) *Naturally, 2019 Closes with Thousands of 10-Inch Pulsing "Penis Fish" Stranded on a California Beach. Bay Nature.* Retrieved from: https://baynature.org/201 9/12/10/naturally-2019-closes-with-thousands-of-10-inch-pulsing-penis-fish-stranded-on-a-california-beach/.

Paton TA, Baker AJ, Groth JG, Barrowclough GF (2003) RAG-1 sequences resolve phylogenetic relationships within charadriiform birds. *Molecular Phylogenetics and Evolution* 29: 268–278.

Pecon-Slattery J, O'Brien SJ (1998) Patterns of Y and X chromosome DNA sequence divergence during the Felidae radiation. *Genetics* 148: 1245–1255.

Platnick NI (2010) *Theridiidae. World Spider Catalog, Version 11.0.* American Museum of Natural History, New York, NY.

Pons J-M, Hassanin A, Crochet P-A (2005) Phylogenetic relationships within the Laridae (Charadriiformes: Aves) inferred from mitochondrial markers. *Molecular Phylogenetics and Evolution* 37: 686–699.

Poole KG (2003) A review of the Canada lynx, *Lynx canadensis*, in Canada. *The Canadian Field-Naturalist* 117(3): 360–376.

Poore GCB (2002) *Superorder: Peracarida Calman, 1905. Crustacea: Malacostraca. Syncarida, Peracarida: Isopoda, Tanaidacea, Mictacea, Thermosbaenacea, Spelaeogriphacea.* CSIRO Publishing, Clayton, Australia. pp. 24–25.

Poulter G (2010). *Becoming Native in a Foreign Land: Sport, Visual Culture, and Identity in Montreal, 1840–1885.* UBC Press, University of British Columbia, Vancouver, British Columbia. p. 33.

Preston-Mafham K (1998) *Spiders: Compact Study Guide and Identifier.* Angus Books, Berkshire Book Co., Sheffield, Massachusetts.

Randall B (2007) *Bushy-Tailed Woodrat Neotoma cinerea.* eNature.com. Retrieved from: http://www.enature.com/fieldguides/detail.asp?recnum=MA0081. June 13, 2021.

Raven PH, Evert RF, Eichhorn SE (2005) *Biology of Plants.* W. H. Freeman, New York, NY.

Resh VH, Cardé RT (2009) *Encyclopedia of Insects.* Academic Press, Cambridge, Massachusetts. p. 722.

Reynolds JW (2016) Earthworms (Oligochaeta: Acanthodrilidae, Lumbricidae, Megascolecidae, Ocnerodrilidae and Sparganophilidae) in the Central California Foothills and Coastal Mountains Ecoregion (6), USA. *Megadrilogica* 21: 73–78.

Rice DW (1998). *Marine Mammals of the World. Systematics and Distribution.* Special Publication Number 4. The Society for Marine Mammalogy, Lawrence, Kansas.

Ripple WJ, Estes JA, Beschta RL, Wilmers CC, Ritchie EG, Hebblewhite M, Berger J, Elmhagen B, Letnic M, Nelson MP, Schmitz OJ, Smith DW, Wallach AD, Wirsing AJ (2014) Status and ecological effects of the world's largest carnivores. *Science* 343: 1241484.

Robbins CS, Brunn B, Zim HS (1983) *Birds of North America; A Guide to Field Identification.* Golden Press, New York, NY.

Rogers-Bennett L (2007) Is climate change contributing to range reductions and localized extinctions in northern (*Haliotis kamtschatkana*) and flat (*Haliotis walallensis*) abalones? *Bulletin of Marine Science* 81(2): 283–296.

Rouhe AC, Sytsma MD (2007) *Feral Swine Action Plan for Oregon*. Portland State University, Portland, Oregon. Retrieved from: http://archives.pdx.edu/ds/psu/4792. June 24, 2021.

Ruppert EE, Fox RS, Barnes RD (2004) *Invertebrate Zoology* (7th ed.). Cengage Learning, Boston, Massachusetts.

Rychel AL, Smith SE, Shimamoto HT, Swalla HT (2006) Evolution and development of the Chordates: Collagen and pharyngeal cartilage. *Mollecular Biology and Evolution* 23(3): 541–549.

Sacks BN, Moore M, Statham MJ, Wittmer HU (2011) A restricted hybrid zone between native and introduced red fox *Vulpes vulpes* populations suggests reproductive barriers and competitive exclusion. *Molecular Ecology* 20: 326–341.

Sacks BN, Statham MJ, Perrine JD, Wisely SM, Aubry KB (2010) North American montane red foxes: Expansion, fragmentation, and the origin of the Sacramento Valley red fox. *Conservation Genetics* 11: 1523–1539.

Sandelin R (2007) Snails and slugs of the Pacific Lowlands. In: *A Field Guide to the Lowland Northwest*. Sky Valley Environments, Snohomish, Washington.

Sandilands Al (2011). *Birds of Ontario: Habitat Requirements, Limiting Factors, and Status: Volume 1–Nonpasserines: Loons through Cranes*. University of British Columbia Press, Vancouver, British Columbia, Canada. p. 171.

Scrafford MA, Boyce MS (2018) Temporal patterns of wolverine (*Gulo gulo luscus*) foraging in the boreal forest. *Journal of Mammalogy* 99(3): 693–701.

Shelley RM (2002) Annotated Checklist of the Millipedes of California (Arthropoda: Diplopoda). *Monographs of the Western North American Naturalist* 1: 90–115.

Sibley DA (2000) *The Sibley Guide to Birds*. Alfred A. Knopf, New York, NY.

Sillero-Zubiri C, Hoffman Michael, MacDonald DW (2004) *Canids: Foxes, Wolves, Jackals, and Dogs: Status Survey and Conservation Action Plan*. IUCN, Gland, Switzerland and Cambridge, UK. p. 95.

Simmons E (2019) *The Importance of Having Insects*. Bay Nature. May 2019.

Smith CC, Haglund TR, Ruiz M, Fisher RN (1993) The status and distribution of freshwater fishes of southern California. *Bulletin of the Southern California Academy of Sciences* 92(3): 101–167.

South A (1992) *Terrestrial Slugs. Biology, Ecology and Control*. Chapman and Hall, London, UK. p. 428.

SpiderID (2018) *Spiders in Washington*. SpiderID, Chicago, Illinois. Retrieved from: https://spiderid.com/locations/united-states/washington/. July 7, 2021.

Statham MJ, Sacks BN, Aubry KB, Perrine JD, Wisely SM (2012) The origin of recently established red fox populations in the United States: Translocations or natural range expansions? *Journal of Mammalogy* 93(1): 52–65.

Stebbins RC (1954) Natural history of the salamanders of the plethodontid genus *Ensatina*. *University of California Publications in Zoology* 54: 47–124.

Stebbins RC (1959) *Reptiles and Amphibians of the San Francisco Bay Region*. University of California Press, Berkeley and Los Angeles.

Stebbins RC (2003) *A Field Guide to Western Reptiles and Amphibians* (3rd ed.). The Peterson Field Guide Series. Houghton Mifflin Company, New York, NY.

Stebbins TD, Eernisse DJ (2009) Chitons (Mollusca: Polyplacophora) known from benthic monitoring programs in the Southern California Bight. *The Festivus, (Special Issue)* 41(6): 53–100 (with errata).

Stebbins RC, McGinis SM (2012) *Field Guide to Amphibians and Reptiles of California: Revised Edition (California Natural History Guides)*. University of California Press, Berkeley, California.

Steppan SJ, Akhverdyan MR, Lyapunova EA, Fraser DG, Vorontsov NN, Hoffman RS, Braun MJ (1999) Molecular phylogeny of the marmots (Rodentia: Sciuridae): tests of evolutionary and biogeographic hypotheses. *Systematic Biology* 48(4): 715–734.

Sullivan PT (1983) A Preliminary Study of Historic and Recent Reports of Grizzly Bears in the North Cascades Area of Washington. Washington Department of Game, Olympia, Washington.

Sunset (1995) *Sunset Western Garden Book; 40th Anniversary Edition*. June 1995. Sunset Publishing Corporation, Menlo Park, CA. pp. 28–31.

Thompson TE (2009) Feeding in nudibranch larvae. *Journal of the Marine Biological Association of the United Kingdom* 38(2): 239–248.

Tighe D (2019) *Meet the Bay's Incredible Swimming Worms*. Bay Nature. Retrieved from: https://baynature.org/2019/08/06/meet-the-bays-incredible-swimming-worms/.

Tomasik E, Cook JA (2005) Mitochondrial phylogeography and conservation genetics of wolverine (*Gulo gulo*) of Northwestern North America. *Journal of Mammalogy* 86(2): 386–396.

Tumlison R (1987) Felis lynx. *Mammalian Species* 269: 1–8.

UCANR (2021) *California Snails and Slugs, Division of Agriculture and Natural Resources*. University of California, Riverside, California. Retrieved from: https://ucanr.edu/sites/CalSnailsandSlugs/. March 13, 2021.

Ueda K-I, Agarwal RG (ND) *California sea slugs - Nudibranchs (and other marine Heterobranchia) of California*. iNaturalist. Retrieved from: https://www.inaturalist.org/guides/40.

UPS (ND) *Seattle Parakeets*. University of Puget Sound, Washington. Retrieved from: https://www.pugetsound.edu/academics/academic-resources/slater-museum/biodiversity-resources/birds/images/land-birds/seattle-parakeets/. June 24, 2021.

USDA (2020) *History of Feral Swine in the Americas*. Animal and Plant Health Inspection Service, US Department of Agriculture, Washington DC. Retrieved from: https://www.aphis.usda.gov/aphis/ourfocus/wildlifedamage/operational-activities/feral-swine/sa-sf-history. June 24, 2021.

USFWS (2013) *Draft Grizzly Bear Restoration Plan / Environmental Impact Statement. North Cascades Ecosystem*. US Fish and Wildlife Service/National Park Service. Retrieved from: https://parkplanning.nps.gov/document.cfm?parkID=327&projectID=44144&documentID=77025. June 15, 2021.

USFWS (2021) *Grays Harbor, National Wildlife Refuge, Washington*. US Fish and Wildlife Service, Washington DC. Retrieved from: https://www.fws.gov/refuge/grays_harbor/. June 28, 2021.

USGS 2780 (ND) Retrieved from: https://nas.er.usgs.gov/queries/FactSheet.aspx?SpeciesID=2780.

UWCAS (ND) *Mammals of Washington*. The Burke Museum, University of Washington, College of Arts and Sciences, Seattle, Washington. Retrieved from: https://www.burkemuseum.org/collections-and-research/biology/mammalogy/mamwash/carnivora.php#Mink, php#Striped_Skunk, php#Western_Spotted_Skunk, php#steller_Sea_Lion. June 22, 2021.

van Tuinen M, Waterhouse DM, Dyke GJ (2004) Avian molecular systematics on the rebound: A fresh look at modern shorebird phylogenetic relationships. *Journal of Avian Biology* 35: 191–194.

Vladykov VD, Follett WI (1965) *Lampetra richardsoni*, a new nonparasitic species of lamprey (Petromyzonidae) from Western North America. *Journal of the Fisheries Research Board of Canada* 22(1): 139–158.

Wagner D (2021) *No Cougar at Discovery Park, But WDFW Says Big Cats Are Closer Than You Might Think.* KIRO News, Seattle, Washington. Retrieved from: https://www.kiro7.com/news/local/no-cougar-discovery-park-wdfw-says-big-cats-are-closer-than-you-might-think/XAD7ZYFOONGI7PRQKFQAAFRNHM/. June 16, 2021.

Wallach AD, Johnson CN, Ritchie EG, O'Neill AJ (2010) Predator control promotes invasive dominated ecological states. *Ecology Letters* 13: 1008–1018.

Ward WW, Cormier MJ (1978) Energy transfer via protein-protein interaction in *Renilla* bioluminescence. *Photochemistry and Photobiology* 27(4): 389–396.

Ward RMP, Krebs CJ (1985) Behavioural responses of lynx to declining snowshoe hare abundance. *Canadian Journal of Zoology* 63(12): 2817–2824.

Washington Native Plant Society (2018) *Rare Plants*. Washington Native Plant Society, Seattle, Washington. Retrieved from: https://www.wnps.org/rare-native-plants. July 11, 2021.

Watkins B (ND) *California Marine Flatworms*. California Diving. Retrieved from: https://cadivingnews.com/california-marine-flatworms/.

Wayne RK, Geffen E, Girman DJ, Koepfli KP, Lau LM, Marshall CR (1997) Molecular systematics of the Canidae. *Systematic Biology* 46(4): 622–653.

WDFW (2015) *State Wildlife Action Plan Update, Appendix A-5, Species of Greatest Conservation Need Fact Sheets, Invertebrates.* Washington Department of Fish and Wildlife, Tacoma, Washington. Retrieved from: https://wdfw.wa.gov/sites/default/files/publications/01742/14_A5_Invertebrates.pdf. July 7, 2021.

WDFW (2019) *State Listed Species*. Washington Department of Fish and Wildlife, Tacoma, Washington. Retrieved from: https://wdfw.wa.gov/sites/default/files/2019-06/threatened-and-endangered-species-list.pdf; June 30, 2021.

WDFW (2021a) *Species in Washington*. Washington Department of Fish & Wildlife, Tacoma, Washington. Retrieved from: https://wdfw.wa.gov/species-habitat/species/. June 23, 2021.

WDFW (2021b) *Razor Clam Seasons and Beaches*. Washington Department of Fish & Wildlife, Tacoma, Washington. Retrieved from: https://wdfw.wa.gov/fishing/shellfishing-regulations/razor-clams. July 7, 2021.

WDFW (2021c) *Crab Indentification and Soft-Shell Crab*. Washington Department of Fish & Wildlife, Tacoma, Washington. Retrieved from: https://wdfw.wa.gov/fishing/basics/crab; July 11, 2021.

WDFW (2021d) *Fishing for Crayfish*. Washington Department of Fish & Wildlife, Tacoma, Washington. Retrieved from: https://wdfw.wa.gov/fishing/basics/crayfish; July 11, 2021.

WDFW (2021e) *Aquatic Invasive Species*. Washington Department of Fish & Wildlife, Tacoma, Washington. Retrieved from: https://wdfw.wa.gov/species-habitat/species. July 11, 2021.

WDFW (NDa) *Living with Wildlife: Tree Squirrels*. Washington Department of Fish and Wildlife, Tacoma, Washington. Retrieved from: https://wdfw.wa.gov/species-habitats/living/species-facts/tree-squirrels; June 12, 2021.

WDFW (NDb) *Living with Wildlife: Cougar (Puma Concolor)*. Washington Department of Fish and Wildlife, Tacoma, Washington. Retrieved from: https://wdfw.wa.gov/species-habitats/species/puma-concolor#living. June 18, 2021.

Weinstein MS (1977) Hares, lynx, and trappers. *The American Naturalist* 111(980): 806–808.

Whittaker RH (1969) New concepts of kingdoms or organisms. *Science* 163(3863): 150–160.

WHSRN (2008a) WHSRN Sites, Western Hemisphere Shorebird Reserve Network, Manomet, Massachusetts.

WHSRN (2008b) WHSRN: A Strategy for Saving Shorebirds. Western Hemisphere Shorebird Reserve Network, Manomet, Massachusetts.

Williams DB (2021) *Digging the Mighty Geoduck: A History of Puget Sound's 'Boss Clam'.* Encyclopedia of Puget Sound, Puget Sound Institute, University of Washington, Tacoma, Washington. Retrieved from: https://www.eopugetsound.org/ magazine/geoduck-clams. July 7, 2021.

Williams BL, Brodie ED III (2003) Coevolution deadly toxins and predator resistance: Self-assessment of resistance by garter snakes leads to behavioral rejection of toxic newt prey. *Herpetologica* 59: 155–163.

Wilson DE, Mittermeier RA (editors) (2009) *Handbook of the Mammals of the World,* Vol. 1. Carnivora. Lynx Ediciones, Barcelona, Spain. pp. 50–658.

Winters C (2016) *The Slowest Invasion: Non-Native Snails Take Over the Northwest.* Everett Herald and Sound Publishing, Inc., Everett, Washington. Retrieved from: https://www.heraldnet.com/news/the-slowest-invasion-non-native-snails-take-over-the-northwest/. July 8, 2021.

Wolcott J (1982) Big frogs, small pond. *Texas Monthly* 10(1): 120.

Woodburne MO (2004) *Late Cretaceous and Cenozoic Mammals of North America: Biostratigraphy and Geochronology.* Columbia University Press, New York, NY.

Woodward SL, Quinn JA (2011) *Encyclopedia of Invasive Species: From Africanized Honey Bees to Zebra Mussels.* ABC-CLIO, Santa Barbara, CA.

Wozencraft WC (2005) Order carnivora. In: Wilson DE, Reeder DM (editors). *Mammal Species of the World: A Taxonomic and Geographic Reference* (3rd ed.), pp. 624–628. Johns Hopkins University Press, Baltimore, MD.

WPZ (ND) *Pacific Martens.* Olympic Marten Project, Woodland Park Zoo, Seattle, Washington. Retrieved from: https://www.zoo.org/martens. June 16, 2021.

WSDOH (ND) Spiders. Washington State Department of Health, Tacoma, Washington. Retrieved from: https://www.doh.wa.gov/communityandenvironment/pests/spiders# Yellow-Sac-Spider. July 7, 2021.

WSU (ND) *Northern Scorpion.* Department of Entomology, Washington State University, Pullman, Washington. Retrieved from: https://entomology.wsu.edu/outreach/bug-info/northern -scorpion/; July 7, 2021.

Yocom CF (1950) Fox squirrels in Asotin County, Washington. *The Murrelet* 31: 34.

Zahran HH (1999) Rhizobium-legume symbiosis and nitrogen fixation under severe conditions and in an arid climate. *Microbiology and Molecular Biology Reviews* 63(4): 968–989.

Zakon HH (2013) Adaptive Evolution of Voltage-Gated Sodium Channels: The First 800 million Years. In Striedter GF, Avise JC, Ayala FJ (editors). *In the Light of Evolution; Volume VI: Brain and Behavior*, pp. 32–34. *The National Academies Press*, Washington, DC.

9 Protecting and Restoring Puget Sound

The Puget Sound is a highly desirable place to live, work, and visit. It has beautiful mountains, water features, forests, and shorelines. However, the very characteristics that are so desirable are also vulnerable to development and overuse. In the last 150 years, in particular, human activities have eroded some of that beauty.

Many of the changes are unavoidable when the population grows as it has in the region. Massive amounts of building and development have occurred for housing, industry, and recreation. Shores have been armored to prevent erosion. Rivers have been contained within their beds to reduce the threats of flooding. Pollutants have seeped into the Sound from sewage systems (or simple raw sewage), industrial processes, agricultural activities, and runoff that carries substances into the Sound.

Most importantly, human activities around the world are causing the Earth to warm at an increasing rate. As the ice caps melt, sea levels rise, rain patterns change, the temperature of bodies of water are increasing, freshwater becomes less available, and changes occur to many other systems. All of this change will stress the plants and animals around the world and in the Sound. Global warming and rising ocean levels are likely to threaten even more wetlands.

Fortunately, in the last few decades, these threats to the Sound and to all of us have been increasingly recognized and efforts have begun to maintain and even restore natural areas, such as wetlands, marshes and salt and mud flats. New laws, such as the Clean Water Act of 1972 and others, have helped. Yet, much remains to be done. But the work will not be easy. First, the remaining natural areas, plants, and animals need to be protected and maintained. New efforts can be made to better manage further development, value the natural areas, and control pollution. Second, areas can be restored to something close to their natural state (Figure 9.1).

Wetlands can be restored along the shore. Rivers can be made more healthy and less polluted. Industrial sites can be cleaned up. In the Puget Sound region, these efforts are underway. There have been successes and challenges remain. Here we will review some of that progress.

RESTORATION

Ecosystems, such as tidal wetlands, are complicated collections of plants, animals, water movement, and chemistry that developed over millions of years (Brush, 2012). It is hard to imagine that they could be recreated by simply

DOI: 10.1201/9780429487439-9

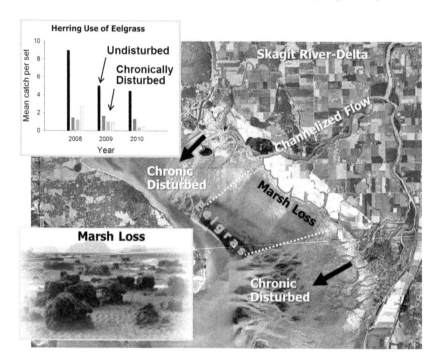

FIGURE 9.1   Degradation of coastal wetlands (Photograph courtesy of US Geological Survey).

mixing back the individual parts. In fact, the complete roster of organisms is rarely known. In many cases, the process is further complicated by the fragmentation of habitat that remains. The limited communication between those pockets hinders cross-fertilization and future diversity. Thus, the restoration of an ecosystem is also quite complicated. For example, how can one measure the functioning of an ecosystem?

Several studies have tried to shed light on restoration. Zedler and Lindig-Cisnerow (2000) asserted that equivalent structure in an ecosystem does not mean equivalent function. The relationships are complicated. Mossman et al. (2012) compared deliberately restored and accidentally restored sites with natural saltmarshes in the United Kingdom. They found that halophytic species adapted rapidly to the managed marshes, but the composition of the communities was still different than the natural communities, and the oxygen and humidity levels might have been lower in the managed marshes. Interestingly, the accidentally realigned marshes were quite similar to the natural marshes. They concluded that we might need to adjust the standards we use to judge marshes. Moreno-Mateos et al. (2020) suggested that the focus must be on very long-term (hundreds to thousands of years) programs to re-establish the habitat. An effort on this time scale gives an opportunity to adjust more effectively. They also suggest a whole-genome sequencing approach to determine what functions are needed in the habitat. This is an intriguing strategy, but to succeed, it would need extraordinary patience and funding. Ruiz-Jaen and Aide (2005) describe ways to measure progress in restorations. They point out that most studies evaluate diversity, vegetation structure, and ecological processes. They also suggested two extra measures: adding variables within each of these three measures that clearly relate to ecosystem functioning and establishing two reference sites to better understand variations in the ecosystems.

Defining appropriate metrics remains an issue for restorations. Clearly, these must be determined for each specific environment. For example, Baggett et al. (2015) offered a set of standardized measures to monitor the restoration of an oyster habitat. They were of two types. First, they proposed four universal metrics: reef areal dimensions, reef height, oyster density, and oyster size-frequency distribution). Second, they added three environmental variables: water temperature, salinity, and dissolved oxygen). They also included several other metrics to monitor the prerestoration and postrestoration efforts. Adoption of a standard set of metrics would allow comparisons of projects and the ability to gauge real progress.

One interesting question is how much of a habitat needs to be restored to the natural condition. Roni et al. (2010) used mathematical modeling to estimate the effects of restoring watersheds on the populations of coho (*Oncorhynchus kisutch*) and steelhead (*O. mykiss*) salmon. They choose two scenarios of restoration: complete restoration and the typical restoration in the Puget Sound region (about 8% of total). In these cases, they determined the percentage of floodplain and in-channel habitat that needed to be restored to allow for a 25% increase in smolt production. That number was 20% of the habitat. However, there was considerable variation in the results and that variation would require 25% of the habitat to be restored to be 95% certain of the desired result. In other words, much more restoration is likely needed than is typically done to achieve an acceptable result.

Another important issue in Puget Sound restorations is the armoring of shorelines. Armoring consists of riprap, bulkheads, and other material that is emplaced to protect against beach erosion, prevent flooding, and protect infrastructure. In fact, much of the Sound shoreline is armored. However, armoring also has undesirable side effects. It also degrades habitat for fish and invertebrates, disrupts the connections between the marine and terrestrial communities, and reduces the overall health of the coast. Restoring wetlands would require those structures to be removed. Even then, could the effects be reversed? Lee et al. (2018) asked just that. They examined six restored sites for five measures (i.e., % wrack cover, % saltmarsh cover, number of logs, and abundance and richness of macroinvertebrates) to answer the question (Figure 9.2). All of the measures were positive in fairly short order (less than one year). By these measures, removing the armoring is a beneficial step in the restoration of habitat.

WETLANDS AND ESTUARIES

Approximately 74% of the vegetated tidal wetlands have been lost from the Puget Sound region (Simenstad et al., 2011). The typical history is that a dike

FIGURE 9.2 Surveying for invertebrates. A USGS biological science technician collects the contents of an invertebrate fallout trap at the Nisqually River Delta, Washington (Photograph courtesy of Sierra Blakely, US Geological Survey).

was built to separate the wetland from the tides. The result was to change the ecosystem from a saltwater to a freshwater environment, and this was accompanied by a dramatic change in the plants and animals in the area. The estuaries have also been injured by the damming of rivers, levees that prevent water from reaching its historic flood plains, and pollution and sediments (Figure 9.1).

Development of wetlands has a number of problems. It disrupts the flow of water and sediment and that adds stress so that some species might be lost. The marine community loses communication with the terrestrial community. Although they are not always as noticeable as wetlands, the transition zones that exist between the wetlands and the terrestrial areas are also very important to the health of the wetlands. Some species need both habitats. Development often fragments natural habitats so that they become isolated from one another. The remaining wetlands may be cut off from other wetlands, and so, they are less likely to be repopulated by native species from the established wetlands. The loss of biodiversity might lead to a less-complex community, more opportunities for invasive species to take root, and less-nitrogen accumulation. Small urban salt marshes are highly vulnerable to exotic species.

Restoration of wetlands and salt marshes takes time. Erfanzadeh et al. (2010) studied the stages of restoration in a salt marsh. They examined seed length, width, and mass to monitor dispersal, and Ellenberg moisture, salinity, and nutrient indices to indicate site suitability. Species appropriate for a salt marsh returned in less than five years. Seed availability was the limiting factor. It was more important than salt tolerance of nutrient limitations. The first plants to arrive produced large numbers of seeds, and those seeds were shorter and lighter. Those characteristics probably helped them to spread more quickly than other species.

Drexler et al. (2019) also compared a restored salt marsh to a natural salt marsh in the southern Puget Sound. Their goal was to determine how well the restored marsh sequestered carbon. This major function of wetlands is becoming even more important with climate change. Core samples from the salt flats were examined for bulk density, loss of ignition, organic carbon, and lead. They found that a restored salt marsh quickly accumulates carbon and can achieve a level similar to that of established salt marshes. This is good news for restorations.

Water processes are sometimes overlooked in planning restorations, and that oversight can present challenges that can be avoided, such as flooding, poor water quality, and erosion. To help remedy this problem, Khangaonkar and Yang (2011) developed a high-resolution circulation and transport model of the Puget Sound. The model describes water quality and circulation in tidal channels, the regular wetting and drying of the tide flats, and sediment transport, and is intended to aid in developing wetland restoration projects.

Cereghino (2015) summarized the thoughts of a number of experts on estuary restoration and distilled them into six key ingredients for a successful project. The team must have sufficient labor. Timely funding is critical to any project. The team must be experienced enough to know how the project will turn out. They must have access to the land. The community must support the project and understand its risks. Regulatory agencies must be satisfied.

Tidal flow is a key element of wetlands restoration, and it is important to determine the proper number of channels to support the transformation and maintenance of the wetlands. Hood (2015) examined restored and reference wetlands in the Puget Sound and Columbia River. He found that there were fivefold more openings in the reference than in the restored sites. The number of outlets is only one factor. How many breeches should be made in the existing dikes and how should they be spaced? Hood suggests that there should be two that face downstream, but others can be placed randomly. Yang et al. (2010) studied the restoration of a 160-acre parcel in the Stillaguamish River estuary. They developed a three-dimensional hydrodynamic model that accounts for the tides, currents, and salinity of the estuary. When they compared the model to the site, they found that the dike removal would be appropriate for the project. The restoration would achieve the desired tidal flushing. Khangaonkar et al. (2017) developed a model of the Swinomish Federal Navigation Channel that connects Skagit and Padilla Bays. Restoration projects there seek to ensure navigation by repairing the dikes that have enabled navigation. They want to control sedimentation without impacting salmon habitat. The model showed that the restoration might not achieve its goals. Thus, the model has provided valuable information to planners about the unintended consequences of the planned actions.

DREDGING AND FILLING

Dredging projects are common in the rivers, estuaries, and Puget Sound itself. Welch et al. (2016) reviewed many of those projects. Dredging and filling have both beneficial and detrimental possibilities. In the past, dredging is important to deepen channels for shipping. For many years, dredged material was considered waste to be disposed of. It was used to provide material to fill areas for development. However, those applications sometimes resulted in the degradation and destruction of habitat. In other cases, the contaminated material was moved to sensitive areas.

In recent years, sediment has come to be seen as a valuable resource that is essential for keeping the Estuary healthy. Material from dredging can be used to fortify beaches and replace sediment lost from wetlands (Carse and Lewis, 2020). Sediment is continually washed out to sea. New sediment, either from natural sources or dredging, can help to maintain a healthy balance of material lost and gained (Milligan and Holmes, 2017). The Stillaguamish Flood Control District provides an example of how dredging and filling can be put to multiple beneficial purposes in an estuary (Fuller, 2015). They dredged the Old Stilly channel to improve drainage on farmland and used the removed material to add sediment to the tidal marshes in the estuary. This was the first large use of dredged material for an estuary project.

Michalsen and Brown (2019) reported on the process and testing that went into a study of the feasibility of placing dredged material into Skagit Bay. The US Army Corp of Engineers is looking for alternatives to the current two deep-water placement sites for dredged material.

Dredging itself stirs up sediment that drifts according to the winds and waves, turbidity, characteristics of the sediments and more (Capello et al., 2010). By better understanding how sediments move, researchers can help others to minimize the detrimental effects of dredging on marine life (Fraser et al., 2017). Dredging increases the amount of suspended sediment and reduces the amount of light under the water. That, in turn, affects marine organisms that depend on photosynthesis. The suspended material also interferes with feeding and other activities. Oysters, for example, are very sensitive to dredging (Wilber and Clarke, 2010). The sediment can physically cover them when it settles over them, and that will interfere with their filter feeding. Finally, oyster larvae need clean hard surfaces to attach to. Even small amounts of loose sediment disrupt the attachment and damage the oyster population.

The disposal of dredged material also disrupts the ecosystem. Bolam and Rees (2003) reviewed the literature to determine the effects on animals in those areas that are subjected to filling. They compared the macrofauna in communities that were environmentally stressed to those in more stable environments. They conclude that the stressed communities recover more rapidly than the more stable communities.

The dredged material might also be contaminated. Many pollutants bind easily to sediments and accumulate at the bottom of the Sound. Sediment that is to be used to restore wetlands must be tested for levels of contaminants to make sure it is safe to distribute. Several options are available for dealing with contaminated sediments under water (Palermo et al., 2000). As the name implies, confined or diked disposal involves placing dredged material within a diked facility that is near the shore or on land. In subaqueous capping, contaminated material is placed in an open water site and then covered by a cap of clean material, such as concrete, that will isolate the contaminated material. There are variations on this model. In level bottom capping, the contaminated material is mounded up on the bottom and covered with a cap. Contained aquatic disposal adds berms or places the material in a depression to prevent lateral dispersion.

The Dredged Material Management Program manages the dredging and disposal of material in the Puget Sound. This interagency program brings together the US Army Corps of Engineers, US Environmental Protection Agency, Washington State Department of Ecology and Washington State Department of Natural Resources. The Puget Sound Dredged Disposal Analysis Study identified acceptable disposal sites and developed plans to monitor the processes. With those items ready, the program transitioned to the Dredged Material Management Program for implementation. Kendall et al. (2003) reviewed the process that resulted in the Program.

PRAIRIES AND LANDFILLS

Prairies need restoration too. One might not associate prairies with the Puget Sound, but the land just south of the Sound features a great deal of prairie. Rook et al. (2011) tested restoration treatments in the south Puget Sound lowlands by looking at the native and invasive plant species. Their results were very interesting. Biodiversity among plants was greater after a prescribed fire. Herbicides reduced the diversity. Proximity to the invasive Scotch broom (*Cytisus scoparius*) reduced the diversity of native species. Those findings will help planners decide which strategy to use in restoration efforts.

One commonly used strategy for encouraging biodiversity has been topographic heterogeneity. Changing the surface adds different niches for different plants to take hold. Landfills are one place that this strategy has been applied. They often need help to restore them as functioning ecosystems. Landfills are typically capped with material that sequesters the waste material, and that cap presents problems to restorations. Their permeability is intentionally poor, and they lack organic material and normal soil fauna. Ewing (2002) conducted an experiment in which mounds were constructed on landfill sites. The mounds consisted of sandy loam, compost, and fertilizer. Prairie plants were added to the mounds. After three years, the effects on weeds were mixed. Some plants (e.g., *Eriophyllum lanatum, Festuca idahoensis, Aster curtus*) thrived. Others (e.g., *Potentilla pacifica, Carex inops*) did not. He concluded that mounding would be one beneficial component of a more comprehensive overall plan.

The Union Bay Natural Area covers 0.3 km^2 and is one of Seattle's largest green spaces. The Area has been improved by 35 restoration projects (Howell and Hough-Snee, 2009). It wasn't always what it is today. Historically, the area was a marsh where the Ravenna, Kincaid, and Yesler Creeks flowed into Lake Washington. It was used as a landfill for maritime and city waste. The large amounts of trash caused the peat to subside. In 1966, the dump was closed, capped with clay, and planted with grass. Some native trees began to grow on the edges, but the cap prevented most native plants from taking hold. As a result, a number of invasive species took over the area. In the late 20th century, the

University of Washington began to actively restore the area and use it to train students in restoration methods.

WILDLIFE

Over the last 150 years, the Puget Sound region has grown into a large metropolitan area. Humans love to visit forests and even live near them. The beauty of the natural environment is a great draw. However, this often puts humans and wildlife on a collision course. Unfortunately, human activities can be detrimental to native species. Even non-motorized activities can disrupt normal routines, affect feeding, and displace animals from their habitat. Much of the traditional habitat for animals has been lost, and what remains can be separated by development and highways.

Wild animals try to adapt by various means. Some come to depend on food left by humans. That could include trash, fruit trees, and even domesticated pets. Mammalian carnivores generally need large ranges and are particularly sensitive to habitat loss and fragmentation. Interestingly, some large carnivores have managed to live in the region, including bobcats (*Lynx rufus*), coyotes (*Canis latrans*), and gray foxes (*Urocyon cinereoargenteus*). Other animals have adapted to living in close contact with human. Coyotes and raccoons (*Procyon lotor*) are among the most successful. They tend to be less particular and will eat almost anything. Coyotes have moved into almost every conceivable habitat in Washington, from open ranch country to densely forested areas to the downtown waterfront (WDFW, 2021). Despite human encroachment and efforts to eliminate them, they have maintained and even increase them in some areas.

Breck et al. (2019) studied the behavior of coyotes from rural and urban areas and made some interesting observations. Urban coyotes were bolder and more exploratory than their country cousins, but there was considerable variation in both groups. They speculated that the rural coyotes are hunted more often and the urban coyotes are not and might even be rewarded. Those behaviors might also lead the coyotes to attack pets or even people.

Seattle is a major city, but it also has more than 400 parks, 130,000 trees, and almost 140 km of shoreline, and that provides habitat for lots of wildlife (Mongrain, 2021). The Seattle Urban Carnivore Project is part of the Seattle Urban Carnivores Project at the Woodland Park Zoo and Seattle University. They use cameras and community reports to track black bears, bobcats, cougars, opossums, raccoons, river otters, and red foxes. Their Carnivore Spotter webpage has documented sightings throughout the Puget Sound region. Within the first two months of its operation, Seattle residents had recorded 2200 observations. Amazingly, all have been reported even in highly developed areas (Carnivore, 2021).

The diets of animals in urban areas vary. Larson et al. (2015) studied bobcats, coyotes, and gray foxes in the urban areas of Southern California. By studying their scat, they found that coyotes ate a broad range of foods, including mammals, fruits, seeds, birds, and invertebrates. Cats were a favorite item. Foxes had a diet similar to coyotes, but foxes rarely take cats. They do very well in urban areas, especially where there are few coyotes. Bobcats mostly eat rodents and rabbits, but will also eat invertebrates and birds. They rarely invade trash cans or attacked cats. Smith et al. (2016) studied raccoons (*Procyon lotot*) and opossums (*Didelphis virginiana*) in Seattle and North Tacoma. Raccoons seemed to prefer human foods to natural foods. Opossums tended to stay in parks and so relied on natural foods. Neither group used marine resources for food.

SPECIFIC RESTORATION PROJECTS

Multiple entities have undertaken projects to restore wetland habitat around the Puget Sound, and more are planned. Several have had notable success. Cereghino (2015) described some of the progress to that date. Christie et al. (2018) expanded that strategy. More has been achieved since.

NISQUALLY NATIONAL WILDLIFE REFUGE

The Nisqually River has a delta that is controlled by sedimentation from the River and erosion by the currents (Takesue and Swarzenski, 2011). For more than 100 years, dikes and levees were built on the salt marshes, and the land was used for agriculture. After being farmed for over 100 years, the area was turned into the Refuge in 1974 (Hodgson et al., 2016). One of the first steps in restoration was breaching an 8-km dike to allow Sound water to enter 3 of the 15.7 km^2 of the Refuge. This step allowed normal tidal action and other biological processes to return to the area. The original goal was to establish habitat and nesting areas for waterfowl and migrating birds, but the Refuge now provides habitat for more than 300 species. The work on the Refuge continues under a 15-year plan that is managed by the Department of Fish and Wildlife

SALMON AND SNOW CREEK ESTUARY

Salmon and Snow Creeks meet to form Discovery Bay. Waste from the lumber industry and other development prevented tidal flow into the Bay and degraded it as habitat. The tidal areas have been filled and pollutants continue to leach into the water. The North Olympic Salmon Coalition restored nearly 45000 km^2 of the area by removing 65,000 cubic yards of material. Other proposals for future work are being considered.

SNOHOMISH RIVER RESTORATION

The Snohomish River enters the Sound at Everett. The Snohomish River drains 48 km^2 of the western Cascades. Its estuary comprised nearly 52 km^2 of marshes, islands, sloughs, and mudflats. The Delta and River have long been habitat for five species of salmon, but the salmon runs today have been reduced by up to ten-fold. Several projects have restored about 5.0 km^2 of the Snohomish River tidal marsh estuary. Today only about 17% of the estuary remains. Channeling of the river with dikes and levees has limited the tidal flooding, and these measures have radically changed the river environment.

SMITH ISLAND RESTORATION

Snohomish County has begun the Smith Island Restoration Project to reestablish tidal wetlands and provide habitat for the Chinook and other salmon species (Smith, 2021). The project involved the building of a new dike to protect farmlands, businesses, and Interstate 5. With that dike in place, they opened up an existing dike to allow tidal flow to return to large areas of wetlands. The newly opened areas provide shelter for juvenile salmon to grow and acclimate to salt water before they leave for the ocean.

QWULOOLT ESTUARY PROJECT

The Tulalip Tribe completed the project to open up more wetlands (Qwuloolt, 2021). Among many actions, they built a new levee to protect infrastructure, removed invasive species and trash, and then breached the north Ebey Slough levee. The tidal action restored a natural habitat for salmon, birds, and other species.

MAJOR CHALLENGES

The number of species of concern is growing in the Puget Sound region (Zier and Gaydos, 2016). The list contained 64 species in 2008, but jumped to 113 in 2011. It is still growing at a rate of 2.6% per year. Some of the increase is due to a better understanding of the animals, but most results from population decline.

INVASIVE SPECIES

The US Fish and Wildlife service (USFWS, 2021) lists a number of marine invasive species. Invasive species arrive by several routes. Some attach directly to ship hulls and drop off or release larvae in distant harbors. Still others are brought intentionally for new populations for fisheries or aquaculture, but the new species is trouble for the new ecosystem. Some organisms actually live on the shells of oysters. When oysters were brought to a new site, the hitchers are inadvertently introduced.

Ballast water is one of the most significant means for invasive species to reach the Puget Sound. Only 16% of marine ecoregions have no reported marine invasions, a number that might be based on underreporting (Molnar et al., 2008). Ships take on water to make the ship more stable while at sea. The crew then discharges the ballast when the ship reaches port. In that way, any plants or animals in the ballast are then released into the new habitat. Each year, more than

1300 ships carrying 15 million cubic meters of ballast water enter the Sound (Cordell et al., 2015).

Washington Administrative Code requires those ships to exchange their ballast water for fresh open ocean water at least 370 km from any US shore and in waters greater than 2000 m. Exceptions are made for ships on ship voyages. Compliance with the rule is quite good but not perfect.

Ballast water loses zooplankton over time. Ships with ballast more than 30 days old usually meet the standard of ten or fewer organisms of 50 mm per cubic meter. Ballast from domestic US waters (e.g., California) had higher densities of non-indigenous zooplankton. Foreign ships tended to have more non-indigenous zooplankton diversity.

Some are listed here (Invasive, 2021). The green crab *(Carcinus maenas)* is native to Europe, but has made it to America. It preys on native clams, oysters, mussels, and crabs. Asian kelp *(Undaria pinnatifida)* is from Japan, China, and Korea and arrived by hull fouling. It outcompetes native species. It has been found on Washington's coasts, but not in Puget Sound yet. The Chinese mitten crab *(Eriocheir sinensis)* is from China and Korea and probably arrived in ballast water. It interferes with fish passage and salvage, water treatment and power plants and more. Three types of invasive tunicates are found in the Puget Sound (Cordell et al., 2013). The clubbed tunicate (*Styela clava*) is from Eastern Asia and likely arrived on the hulls of ships or with imported oysters. Three invasive tunicate species are present in Puget Sound. *Didemnum vexillum* forms bloblike colonies. *Ciona savignyi* forms transparent tubes. The African clawed frog (*Xenopus laevis*) has been found in two watersheds in the Puget Sound. It reproduces rapidly and outcompetes native frogs.

There are also invasive plants. For example, the common reed (Phagmites australis) is found in or near wetlands or other areas with water. It uses rhizomes to spread horizontally and crowd out native grasses. "Killer algae" (*Caulerpa taxifolia)* is from the Mediterranean Sea. It has caused huge problems in California, and officials fear it could be a problem in the Puget Sound. Cordgrass or spartina (*Spartina alterniflora*) has been found in various parts of Puget Sound where it is removed by an eradication team from the Washington Department of Agriculture. Once started, the plant spreads quickly along saltwater shorelines and freshwater areas. It can outcompete most native plant species, reduce marsh biodiversity, and alter ecological functions.

Some invasive species are on land. Scotch broom (*Cytisus scoparius*) is a woody, leguminous shrub originally from Europe and North Africa (Hulting et al., 2008). It was brought to California in the 1850s as an ornamental plant. However, it has spread throughout much of the Northwest. It rapidly colonizes disturbed areas to form dense stands. It outcompetes native plants and causes serious damage to the lumber industry.

Invasive plants have taken over about 94% of Seattle's ~32 km^2 of public land and 20% of the forested areas (Ramsay et al., 2004). The Seattle Urban Nature Project inventoried the public land for habitat types, acreage, plant species, and cover estimates. Those data were used to prioritize restoration efforts. Their findings can be used to evaluate site quality and reference conditions.

Specific Challenges

If any species characterizes the Puget Sound, it is the Chinook salmon. Sadly, the populations of those fish are far lower today that they were (NOAA, 2007). Currently, 22 populations remain. More than 15 Chinook runs have been lost due to loss of their habitat for spawning. Chinook numbers are 10% of their historic levels. In some rivers, those numbers are only 1%.

One aspect of restoring the Sound is simply reestablishing those key species that support other species. For example, eelgrass (*Zostera marina*) is an important component of the nearshore habitat and a key indicator of the health of the ecosystem. Since 2013, Washington has focused on restoring eelgrass sites in the Puget Sound. Gaeckle et al. (2018) evaluated the progress of eelgrass restoration at 15 test sites. They are now following those sites to determine the best stocks, the recovery, and changes to the water chemistry.

Historically, populations of the Olympia oyster (*Ostrea lurida*) were huge, but over harvesting and the loss of habitat have caused them to decline dramatically (Fresh et al., 2011). Although they are no longer harvested, they have not recovered because of habitat loss, silting and pollution and are estimated to be 10% of their original numbers. Bivalves need sediment for their larvae to develop. Unfortunately, over 25% of the Sound's shore is armored, and that has greatly affected the bluff erosion and sediment replenishment. The oyster populations are unlikely to recover unless some areas are reserved for these natural processes. Now efforts are underway to restore the oysters. To help in these efforts, Grossman et al. (2020) examined the process of larval transport. They found that tides, seasons, or combinations were not involved with larval movement. The larvae did seem to be carried into a northward gyre during the initial and mid-ebbtides. The findings will be helpful to those trying to reestablish the oysters.

Forage fish are critical to any ecosystem. These small pelagic fish are the main food source for many larger fish, birds, and mammals. In the Puget Sound, they belong to three main species: surf smelt, Pacific sand lance, and Pacific herring. Their spawning areas are vulnerable to shoreline and beach development (e.g., armoring, marinas, roads). Sediment transport is disrupted. The beach surface becomes hotter where plants are removed.

Sea Level Rise

Sea level rise due to global warming is a serious threat to nearly all coastal areas, and the Puget Sound region is one of those area at risk. The average sea level has already risen, and extreme tides are more common. Estimates were that sea levels will rise 1.7 mm/year in the late 20th century to 3.1 mm/year in the early 21st century. Many low-lying areas will be at greater risk for inundation. Other aspects of climate change (e.g., storm intensity, rainfall, storm surges) will be exacerbated by climate change. Different locations will be affected to greater or lesser degrees, and these factors will make is more difficult to maintain shorelines and native species in the future However, knowing what is likely to happen is critical for planning how best to adapt to those changes.

Local land subsidence may also make things worse, but maps displaying the risk of subsidence are lacking. Those correlated with areas of artificial landfill or Holocene mud deposits. They predicted an area of 125–429 km^2 is at risk of flooding, which is somewhat higher than when only sealevel rise is considered. This more-accurate mapping will help governmental agencies with planning for the future. Subsidence can be caused by a number of factors, including not watering soil, compaction, pumping out groundwater, and removing crude oil (Brandt et al., 2020). The use of metal-based coagulants to bind material together is one possible solution. To test this idea, one study (Stumpner et al., 2018) flooded wetlands with water containing coagulants that would accrete minerals. After 23 months of treatment, sediment samples were examined. Wetlands treated with polyaluminum chloride worked the best, but iron sulfate was also good. Other characteristics of the accreted material were also encouraging. Overall, these coagulants seem to have efficacy in reversing subsidence and storing carbon. By using them, levee failure might be reduced.

Wetlands restoration projects add to the defense of the coast by providing additional areas for increased water levels to be absorbed even before they make contact with the seawalls, levees, and dikes that protect infrastructure and other items.

IMPROVING THE AIR QUALITY

Unfortunately, the air quality in Seattle-Tacoma of some concern. In its State of the Air report, the American Lung Association ranked Seattle-Tacoma 29 of 229 metropolitan areas for high ozone levels, 14 of 216 for 24-hour particulates and 61 of 204 for annual particulate pollution (ALA, 2021). In the last few years, wildfires have also added an enormous amount of particulate matter to the air around Puget Sound. At one point in 2020, Seattle had the worst air in the United States.

Climate change is likely to cause the air quality to become worse in the near future (Mote and Salathé, 2010). Even now, excessive heat and drought have resulted in fires that made the air the worst in the world for a while in 2018. The continuing influx of people will also bring additional air pollution due to the continuing use of fossil fuels.

HOPEFUL SIGNS

PEREGRINE FALCONS

In other cities, Peregrine falcons (*Falco peregrinus*) have thrived by preying on other birds mostly, such as rock doves (*Columba livia*). They nest in nesting in churches, window ledges of skyscrapers, and the towers of suspension bridges. Peregrines were nearly wiped out by pesticides, but they have made a comeback around the Puget Sound sincedichlorodiphenyltrichloroethane or DDT was banned in 1972 (Urban Raptors, 2020). It took nearly 20 years for peregrines to return to Seattle. By 1999, the falcons had recovered sufficiently to be removed from the Endangered Species List. The breeding site at 1201 Third Avenue has been exceptionally successful. Five other breeding sites are known, all on bridges. Of 57 fledged, 11 died quickly. Males seemed to stay in Seattle. Females migrated farther. One nested in central Oregon. In 2000, a nesting pair of per- egrines was found in Tacoma. From 2012 to 2018, 23 birds fledged from a site there (Hartman, 2018).

Bald Eagles

DDT also affected bald eagles (*Haliaeetus leucocephalus*) and greatly reduced their numbers (Encyclopedia, nd). In 1978, they were listed as "threatened" under the Endangered Species Act. However, they have made a good recovery, and they are still protected by the Bald Eagle Protection Law of 1984. The eagles live year-round in the Puget Sound region. They feed on fish and waterfowl and also scavenge. They prefer to nest near large bodies of open water and need a large range for hunting. Continuing threats to the eagles consist of habitat de- gradation and loss of prey. In 2005, there were 840 occupied nests.

Humpback Whales

Humpback whales (*Megaptera novaeangliae*) have also made a strong comeback in the Puget Sound (Puget Sound Express, 2016). Once hunted to near extinction, the whales are now commonly seen in groups of 15–20. Several reasons have been suggested to explain the increase in numbers, but it isn't clear. The in- creased numbers of whales may need more feeding areas, so they have moved back into the Sound. They generally feed on krill and schools of small fish (e.g., sardines, sand lance, anchovy, and herring).

Species Diversity and Stability

The health of an ecosystem can be measured in part by knowing the species diversity. These measures are especially important for maintaining and restoring areas, such as the Puget Sound. Most of these studies focus on species richness or abundance, but those measures alone might not capture all of the complexity of a community. Other efforts have tried to add measure of the phylogenetic and functional aspects of biodiversity. Hultgren et al. (2021) attempted to this with crustaceans at 11 sites in the Puget Sound. They looked at habitat complexity, algal richness, salinity, and urbanization. Interestingly, algal richness was im- portant for diversity. Boulder size was correlated with functional diversity. Efforts to maintain and restore ecosystem are informed by methods to monitor the health of the system. Kershner et al. (2011) considered a set of indicators for an ecosystem, specifically the Puget Sound. Their framework links indicators to policy goals so they can be used successfully by many users.

MAINTAINING AND RESTORING THE SOUND

We cannot wave a magic wand and return the Sound to what it was 10,000 years ago or even 200 years ago. Nearly four million people live and work in the region, and more will arrive in the next 20 years. The problem will get worse as more people need housing, jobs, transportation, and recreation. They will need freshwater and air and create waste.

The nearshore ecosystem fits between the marine and terrestrial territories. It contains birds, shellfish, mammals, fish, and more. In the Puget Sound, it lies along 4000 km of shoreline, estuaries, and marine systems. Humans have made enormous changes to this environment. Fresh et al. (2011) catalogued the changes into four types. The 15 largest river deltas have lost 27% of their area. Over 300 small embayments have been eliminated by development. Shoreline armoring has reduced the influx of sediment to the beaches. Much of the tidal wetlands have been lost, especially in the Puyallup and Duwamish deltas. These have significantly affected the plants and animals of the Sound. For example, several salmon species on the Endangered Species List and killer whales rely on the nearshore ecosystem. Bird numbers are in decline.

Wetlands are critical habitat. Each year tens of thousands of shorebirds and waterfowl use the mudflats and salt ponds to rest and feed as they migrate along the Pacific Flyway. And beyond the needs of birds and other species are the needs of the growing human population in the region. Some balance must be struck between the competing needs of those birds and the people who live in Puget Sound region.

The good news is that there is a far-greater realization of the importance of the environment and the value of wetlands and wild species than ever before. In addition, legislation now aims to protect the environment and to prevent pollution. It will take all of this and more to maintain the environment in the face of additional growth, sea level rise, and global warming.

Parks are valuable additions to cities. Their benefits to the psychological well-being of residents are well documented to reduce stress, depression, and aggression. Nevertheless, parks are often under pressure of development and simple overuse. Even seemingly innocent human activities can be disruptive to animal habitats.

Discovery Park in Seattle contains more than 2.0 km^2 and 19 km of trails. Lev et al. (2020) surveyed visitors to get their impressions of the park. The common responses were "encountering wildlife," "following trails," "walking to destination spots in nature," and "gazing out at the Puget Sound or mountains," and "walking along edges of beach or bluffs."

REFERENCES

ALA (2021) *Seattle-Tacoma*. American Lung Association. Retrieved from: https://www.lung.org/research/sota/city-rankings/msas/seattle-tacoma-wa. April 29, 2021.

Baggett LP, Powers SP, Brumbaugh RD, Coen LD, DeAngelis BM, Greene JK, Hancock BT, Morlock SM, Allen BL, Brietburg DL, Bushek D, Grabowski JH, Grizzle RE, Grosholz ED, La Peyre MK, Luckenbach MW, McGraw KA, Piehler MF, Westby SR, zu Ermgassen PSE (2015) Guidelines for evaluating performance of oyster habitat restoration. *Restoration Ecology* 23: 737–745.

Bolam SG, Rees HL (2003) Minimizing impacts of maintenance dredged material disposal in the coastal environment: A habitat approach. *Environmental Management* 32: 171–188.

Brandt JT, Sneed M, Danskin WR (2020) Detection and measurement of land subsidence and uplift using interferometric synthetic aperture radar, San Diego, California, USA, 2016–2018. *Proceeding of the International Society of Hydrological Sciences* 382: 45–49.

Breck SW, Poessel SA, Mahoney P, Young JK (2019) The intrepid urban coyote: A comparison of bold and exploratory behavior in coyotes from urban and rural environments. *Science Reports* 9: 2104.

Brush S (2012) *The Restoration Project*. Retrieved from: http://therestorationproject.weebly.com/wetland-restoration---puget-sound.html. April 24, 2021.

Capello M, Cutroneo L, Castellano M, Orsi M, Pieracci A, Bertolotto RM, Povero P, Tucci S (2010) Physical and sedimentological characterisation of dredged sediments. *Chemistry and Ecology* 26(suppl. 1): 359–369.

Carnivore (2021) *Seattle Urban Carnivore Project at the Woodland Park Zoo and Seattle University*. Retrieved from: https://www.zoo.org/seattlecarnivores. April 28, 2021.

Carse A, Lewis JA (2020) New horizons for dredging research: The ecology and politics of harbor deepening in the southeastern United States. *WIREs Water* e1485. DOI: 10.1002/wat2.1485.

Cereghino PR (2015) *Recommendations for Accelerating Estuary Restoration in Puget Sound*. Puget Sound Partnership and Washington Department of Fish and Wildlife by NOAA Fisheries Restoration Center, Seattle, WA. p. 20.

Christie P, Fluharty D, Kennard H, Pollnac R, Warren B, Williams T (2018) Policy pivot in Puget Sound: Lessons learned from marine protected areas and tribally-led estuarine restoration. *Ocean & Coastal Management* 163: 72–81.

Cordell JR, Kalata O, Pleus A, Newsom AJ, Strieck K, Gertsen G (2015) *Effectiveness of Ballast Water Exchange in Protecting Puget Sound from Invasive Species*. Washington Department of Fish and Wildlife, Olympia WA. pp. 1–55.

Cordell JR, Levy C, Toft JD (2013) Ecological implications of invasive tunicates associated with artificial structures in Puget Sound, Washington, USA. *Biological Invasions* 6: 1303–1318.

Drexler JZ, Woo I, Fuller CC, Nakai G (2019) Carbon accumulation and vertical accretion in a restored versus historic salt marsh in southern Puget Sound, Washington, United States. *Restoration Ecology* 27: 1117–1127.

Encyclopedia (nd) *Bald Eagles*. Encyclopedia of Puget Sound. Retrieved from: https://www.eopugetsound.org/science-review/10-bald-eagles. April 26, 2021.

Erfanzadeh R, Garbutt A, Petition J, Maelfait JP, Hoffmann M (2010) Factors affecting the success of early salt-marsh colonizers: Seed availability rather than site suitability and dispersal traits. *Plant Ecology* 206: 335–347.

Ewing K (2002) Mounding as a technique for restoration of prairie on a capped landfill in the Puget Sound Lowlands. *Restoration Ecology* 10: 289–296.

Fraser M, Short J, Kendrick G, McLean D, Keesing J, Byrne M, Caley MJ, Clarke D, Davis A, Erftemeijer P, Field S, Gustin-Craig S, Huisman J, Keough M, Lavery P, Masini R, McMahon K, Mengersen K, Rasheed M, Statton J, Stoddart J, Wu P (2017) Effects of dredging on critical ecological processes for marine invertebrates, seagrasses and macroalgae, and the potential for management with environmental windows using Western Australia as a case study. *Ecological Indicators* 78: 229–242.

Fresh K, Dethier M, Simenstad C, Logsdon M, Shipman H, Tanner C, Leschine T, Mumford, Gelfenbaum G, Shuman R, Newton J (2011) *Implications of Observed Anthropogenic Changes to the Nearshore Ecosystems in Puget Sound.* Puget Sound Nearshore Ecosystem Restoration Project. Technical Report 2011-03.

Fuller R (2015) *Beneficial Re-use of Dredged Sediment to Enhance Stillaguamish Tidal Wetlands.* Retrieved from: https://salishsearestoration.org/images/2/2f/Fuller_2014_ stillaguamish_dredged_sediment_reuse.pdf. April 27, 2021.

Gaeckle J, Vavrinec J, Buenau K, Borde AB, Aston L, Thom RM, Shannon J (2018) *Eelgrass (Zostera marina) Restoration in Puget Sound: Restoration Tools, Successes and Challenges.* Salish Sea Ecosystem Conference, Seattle, WA. Retrieved from: https://cedar.wwu.edu/ssec/2018ssec/allsessions/168/. April 26, 2021.

Grossman SK, Grossman EE, Barber JS, Gamble SK, Crosby SC (2020) Distribution and transport of Olympia oyster *Ostrea lurida* larvae in Northern Puget Sound, Washington. *Journal of Shellfish Research* 39: 215–233.

Hartman S (2018) *An Introduction to Tacoma's Peregrine Falcons.* Grit City Magazine. Retrieved from: https://gritcitymag.com/2018/06/an-introduction-to-tacomas-peregrine-falcons/. April 26, 2021.

Hodgson S, Ellings C, Rubin SP, Hayes MC, Duval W, Grossman EE (2016). *2010–2015 Juvenile Fish Ecology in the Nisqually River Delta and Nisqually Reach Aquatic Reserve.* Nisqually Indian Tribe, Department of Natural Resources, Olympia, Washington.

Hood WG (2015) Predicting the number, orientation and spacing of dike breaches for tidal marsh restoration. *Ecological Engineering* 83: 319–327.

Howell J, Hough-Snee N (2009) Learning from a landfill: Ecological restoration and education at Seattle's Union Bay Natural Area. *SER News: The Newsletter of the Society for Ecological Restoration International* 23(2): 4–5.

Hultgren KM, Ossentjuk L, Hendricks K, Serafin A (2021) Crustacean diversity in the Puget Sound: Reconciling species, phylogenetic, and functional diversity. *Marine Biodiversity* 51: 37.

Hulting A, Neff K, Coombs E, Parker R, Miller G, Burrill LC (2008) *Scotch Broom. Biology and Management in the Pacific Northwest.* Oregon State University, Oregon.

Invasive (2021) *Protecting Washington's Environment and Economy from Harmful Invasive Species.* Washington Invasive Species Council. Retrieved from: https://invasivespecies.wa.gov/. April 29, 2021.

Kendall D, Stirling S, Cole-Warner L, Gries T, Vining R, Benson T, Brenner R, Barton J, Hoffman E, Malek J (2003) *Thirteen Year Implementation Retrospective on the Dredged Material Management Program (DMMP) in the Northwest.* Retrieved from: https://www.researchgate.net/profile/Tom-Gries/publication/267821277_Thirteen_ Year_Implementation_Retrospective_on_the_Dredged_Material_Management_ Program_DMMP_in_the_Northwest/links/56d895bb08aebabdb40d1c0f/Thirteen-Year-Implementation-Retrospective-on-the-Dredged-Material-Management-Program-DMMP-in-the-Northwest.pdf. April 27, 2021.

Kershner J, Samhouri JF, James CA, Levin PS (2011) Selecting indicator portfolios for marine species and food webs: A Puget Sound case study. *PLoS ONE* 6(10): e25248.

Khangaonkar T, Nugraha A, Hinton S, Michalsen D, Brown S (2017) Sediment transport into the Swinomish Navigation Channel, Puget Sound—Habitat restoration versus navigation maintenance needs. *Journal of Marine Science and Engineering* 5: 19.

Khangaonkar T, Yang Z (2011) A high-resolution hydrodynamic model of Puget Sound to support nearshore restoration feasibility analysis and design. *Ecological Restoration* 29: 173–184.

Larson RN, Morin DJ, Wierzbowska IA, Crooks KR (2015) Food habits of coyotes, gray foxes, and bobcats in a coastal southern California urban landscape. *Western North American Naturalist* 75(3): 10.

Lee TS, Toft JD, Cordell JR, Dethier MN, Adams JW, Kelly RP (2018) Quantifying the effectiveness of shoreline armoring removal on coastal biota of Puget Sound. *Peer J* 6: e4275. DOI: 10.7717/peerj.4275.

Lev E, Kahn Jr PH, Chen H, Esperum G (2020) Relatively wild urban parks can promote human resilience and flourishing: A case study of Discovery Park, Seattle, Washington. *Frontiers in Sustainable Cities* 2: 2.

Michalsen DR, Brown SH (2019) *Regional Sediment Management: Integrated Solutions for Sediment Related Challenges Feasibility of Nearshore Placement Near the Swinomish Navigation Channel; Puget Sound, Washington.* US Army Corps of Engineers. Retrieved from: https://apps.dtic.mil/sti/pdfs/AD1079678.pdf. April 27, 2021.

Milligan B, Holmes R (2017) Sediment is critical infrastructure for the future of California's Bay-Delta. *Shore & Beach* 85(2).

Molnar JL, Gamboa RL, Revenga C, Spalding MD (2008) Assessing the global threat of invasive species to marine biodiversity. *Frontiers in Ecology and the Environment* 6: 485–492. Retrieved from: https://citeseerx.ist.psu.edu/viewdoc/download?doi=1 0.1.1.909.7219&rep=rep1&type=pdf.

Mongrain R (2021) *The Pacific Northwest–a Haven for Urban Wildlife.* Project Nature. Retrieved from: https://www.projectnaturewa.com/the-pacific-northwest-a-haven-for-urban-wildlife/. April 28, 2021.

Moreno-Mateos D, Alberd A, Morriën E, van der Putten WH, Rodríguez-Uña A, Montoya D (2020) The long-term restoration of ecosystem complexity. *Nature Ecology and Evolution* 4: 676–685.

Mossman HL, Davy AJ, Grant A (2012) Does managed coastal realignment create salt-marshes with 'equivalent biological characteristics' to natural reference sites? *Journal of Applied Ecology* 49: 1446–1456.

Mote PW, Salathé Jr EP (2010) Future climate in the Pacific Northwest. *Climatic Change* 109: 29–50.

NOAA (2007) *Recovery Plan for Puget Sound Chinook Salmon.* National Oceanic and Atmospheric Administration. Retrieved from: https://www.fisheries.noaa.gov/resource/document/recovery-plan-puget-sound-chinook-salmon. April 26, 2021.

Palermo MR, Clausner JE, Channell MG, Averett DE (2000) *Multiuser Disposal Sites (MUDS) for Contaminated Sediments from Puget Sound—Subaqueous Capping and Confined Disposal Alternatives.* ERDC TR-00-3, U.S. Army Engineer Research and Development Center, Vicksburg, MS.

Puget Sound Express (2016) *We're Seeing Unprecedented Numbers of Humpbacks.* Puget Sound Express. Retrieved from: https://www.pugetsoundexpress.com/were-seeing-unprecedented-numbers-of-humpbacks/. April 26, 2021.

Qwuloolt (2021) *Qwullolt Estuary. A project of the Tulalip Tribe.* Retrieved from: https://www.qwuloolt.org/. April 26, 2021.

Ramsay M, Salisbury N, Surbey S (2004) *A Citywide Survey of Habitats on Public Land in Seattle, a Tool for Urban Restoration Planning and Ecological Monitoring*. 16th International Conference, Society for Ecological Restoration, Victoria, Canada.

Roni P, Pess G, Beechie T, Morley S (2010) Estimating changes in Coho salmon and steelhead abundance from watershed restoration: How much restoration is needed to measurably increase smolt production? *North American Journal of Fisheries Management* 30: 1469–1484.

Rook EJ, Fischer DG, Seyferth RD, Kirsch JL, LeRoy CJ, Hamman S (2011) Responses of prairie vegetation to fire, herbicide, and invasive species legacy. *Northwest Science* 85: 288–302.

Ruiz-Jaen MC, Aide TM (2005) Restoration success: How is it being measured? *Restoration Ecology* 13: 569–577.

Simenstad CA, Ramirez M, Burke B, Logsdon L, Shipman H, Tanner C, Toft J, Craig B, Davis C, Fung J, Bloch P, Fresh K, Myers D, Iverson E, Bailey A, Schlenger P, Kiblinger C, Myre P, Gerstel W, MacLennan A (2011) *Historical Change of Puget Sound Shorelines: Puget Sound Nearshore Ecosystem Project Change Analysis*. Puget Sound Nearshore Report No. 2011-01. Washington Department of Fish and Wildlife, Olympia, Washington, and U.S. Army Corps of Engineers, Seattle, Washington.

Smith (2021) *Smith Island Restoration Project*. Snohomish County. Retrieved from: https://snohomishcountywa.gov/1150/Smith-Island-Restoration-Project. April 26, 2021.

Smith M, Wimberger P, Jordan M, Fox-Dobbs K (2016) Using stable isotopes to determine resource partitioning and the role of anthropogenic food sources on urban populations of Seattle and North Tacoma raccoons (*Procyon lotor*) and opossums (*Didelphis virginiana*). *Summer Research* 287. Retrieved from: https://soundideas.pugetsound.edu/summer_research/287.

Stumpner EB, Kraus TEC, Liang YL, Bachand SM, Horwath WR, Bachand PAM (2018) Sediment accretion and carbon storage in constructed wetlands receiving water treated with metal-based coagulants. *Ecological Engineering* 111: 176–185.

Takesue RK, Swarzenski PW (2011) *More Than 100 Years of Background-Level Sedimentary Metals, Nisqually River Delta, South Puget Sound*. U.S. Geological Survey, Washington. p. 13. Retrieved from: https://pubs.usgs.gov/of/2010/1329/.

Urban Raptors (2020) *Seattle Peregrines*. Urban Raptors. Retrieved from: https://urbanraptorconservancy.org/seattle-urban-raptors/falcons/peregrine-falcon/seattles-peregrines/#:~:text=Urban%20History,WaMu%2C%20at%201201%20Third%20Avenue. April 26, 2021.

USFWS (2021) *Invasive Species*. Washington Fish and Wildlife Office. US Fish and Wildlife Service. Retrieved from: https://www.fws.gov/wafwo/articles.cfm?id=149489483. April 29, 2021.

WDFW (2021) Coyote (*Canis latrans*). Washington Department of Fish and Wildlife. Retrieved from: https://wdfw.wa.gov/species-habitats/species/canis-latrans#desc-range. April 28, 2021.

Welch M, Mogren ET, Beeney L (2016) *A Literature Review of the Beneficial Use of Dredged Material and Sediment Management Plans and Strategies*. Portland State University, Hatfield School of Government, Center for Public Service Publications and Reports. Retrieved from: https://pdxscholar.library.pdx.edu/publicservice_pub/34.

Wilber D, Clarke D (2010) Dredging activities and the potential impacts of sediment resuspension and sedimentation on oyster reefs. Proceedings of the Western Dredging Association Technical Conference, June 6–9, 2010, San Juan, Puerto Rico, USA, pp. 61–69.

Yang Z, Sobocinski KL, Heatwole D, Khangaonkar T, Tom R, Fuller R (2010) Hydrodynamic and ecological assessment of nearshore restoration: A modeling study. *Ecological Modelling* 221: 1043–1053.

Zedler JB, Lindig-Cisnerow R (2000) Functional equivalency of restored and natural salt marshes. In: MP Weinstein, DA Kreeger (editors). *Concepts and Controversies in Tidal Marsh Ecology,*pp. 565–582. Springer, New York.

Zier J, Gaydos JK (2016). The growing number of species of concern in the Salish Sea suggests ecosystem decay is outpacing recovery. Proceedings of the 2016 Salish Sea Ecosystem Conference, April 13–15, 2016, Vancouver, BC.

10 Puget Sound in the Future

The Puget Sound that we know today was millions of years in the building. It has been relatively stable since the last Ice Age ended. Sea levels have risen and fallen over that time so that specific areas were either dry or under water. The volcanoes are long extinct. That was also approximately the time when humans arrived. So, for the entire period of human awareness, the Sound has been just as it is today. However, the forces that built the Sound are still at work, and they will eventually change the Sound beyond recognition. Earthquakes and the tectonic migrations they represent are the largest, but not the only forces still at work. Volcanoes, rising sea levels, atmospheric rivers, droughts, landslides, and fires will continue to affect the region. Over time, the Sound will be unrecognizable.

The Sound region is reminded of those forces periodically. The most dramatic reminders are earthquakes. In fact, earthquakes larger than magnitude 6.0 occurred in 1909, 1939, 1946, 1949, 1965, and 2001 (SOEM, nd). The region has also experienced massive megathrust earthquakes. These can reach a magnitude of 9.0 and are caused by a break along the entire subduction zone. They occur about every 500 years. Over a million years—not much time geologically—the region will experience 2000 of these massive earthquakes.

In the future, these forces will change the Puget Sound region. They are examined individually in more detail below. However, it is important to remember that these forces can act at the same time to multiply their combined effects.

GROWTH AND DEVELOPMENT

MORE PEOPLE

The greatest force for change to the Sound over the last thousands of years has been human activities. These have accelerated in the last 200 years since the Europeans arrived. Today, the Seattle-Tacoma metro area has four million people. It is not surprising. The area is extraordinarily beautiful with a mild, if somewhat wet, climate. It has a world-class harbor. It is a highly desirable place to live and work, and thus, the population is likely to continue to grow over the next decades.

By 2050, that number is expected to grow to nearly six million. The additional two million people will need homes, jobs, roads, bridges, water, sewage, electricity, recreation, and more. The natural setting is also fragile, and the pressures of more humans will stress all of those natural elements that are so desirable.

DOI: 10.1201/9780429487439-10

Change began with the arrival of humans. They were slow in the beginning. Native Americans arrived over 10,000 years ago, but they contributed only small changes. The arrival of Europeans and Americans brought huge changes. Wetlands were filled and drained. Rivers were channeled. And a wide variety of pollutants were released into the Bay water. The quality of water and air deteriorated.

FEWER PLANTS AND ANIMALS

Unfortunately, as more people need more room, that leaves less room for plants and animals. The problem is not always less available habitat, but that the patches of remaining habitat are increasingly separated. Some species, especially predators, need a large area to roam, and those larger tracks are fewer as the human population and development increase. In addition, many animals and even plants lose genetic diversity if their range is bounded.

Sadly, that translates into the loss of biodiversity and fewer total animals in many cases. For example, nearly 75% of flying insects have been lost over the last 27 years (Hallmann et al., 2017). Large number of birds are being lost (Loss et al., 2015). In another analysis, Mason et al. (2021) used computer modeling to examine the ranges of birds that are not considered to be threatened. They found that 42% of the birds have wider ranges than expected and 28% have narrower ranges. Finally, amphibians and particularly frogs have suffered enormous losses in recent years for unknown causes (Allentoft and O'Brien, 2010). These losses also have practical implications. Honey bees are critical to the pollination of many of our food crops, and they are in serious decline (Hristov et al., 2020).

Many scientists have begun to wonder if we are witnessing a sixth mass extinction. Five mass extinctions have occurred in Earth's history in which very large numbers of species were lost in geologically short periods of time. The most famous mass extinction killed off the dinosaurs. The causes of these events are unclear. The rate of loss of species today suggests that another event is underway, but this time, the cause is clear. Human activities (e.g., development, habitat loss, and partitioning) are responsible, and the major culprit is climate change.

The Puget Sound region is also experiencing a loss of species. Dunwiddie et al. (2014) examined 80 species of native annual plants on the prairies in the southern part of the Sound. They found that 39% of them do not appear on any recent survey in the region, and another 21% were seen at only one site. Habitat loss is a serious issue, but Feist et al. (2018) noted that toxic stormwater runoff is another cause of concern. They looked at the threat to the coho salmon in the Sound. These fish are considered a sentinel species. Feist et al. compared mortality rates at 51 spanning sites throughout the region to the degree of urbanization in those areas (e.g., traffic rates). Deaths showed a positive correlation with urbanization, and the number of deaths was even greater during months with rain. Thus, many of the spanning sites were adversely affected by the runoff.

At least 9 non-native species of marine plants and 83 species of marine animals have invaded the Sound (Encyclopedia, nd). For example, cordgrass (*Spartina alterniflora*) outcompetes many native species and leads to reduced biodiversity. European green crab *(Carcinus maenas)* is a small, aggressive marine shore crab that damages the native environment. Three tunicate species (*Styela clava, Didemnum vexillum,* and *Ciona savignyi*) grow in large numbers.

Fortunately, in the last few decades, there has been increasing awareness of the importance of the environment. For example, wetlands are needed as a defense against sea level rise and to protect wildlife in the surrounding areas. However, the continuing development of the Puget Sound region will need to be carefully balanced against the needs of the natural environment, and that will require considerable political will.

ATMOSPHERIC RIVERS

The Northwest is well known for its rain. The climate crisis will likely make storms everywhere more intense. In the Northwest, winters will be warmer and wetter, and summers will be warmer and drier (Miller et al., 2018). There will be more snow in the Cascades, but droughts will be longer too.

Atmospheric rivers bring enormous amounts of water vapor to the coasts. Their frequency and duration are greatest on the Oregon–Washington coast (Rutz et al., 2014). These extratropical cyclones yield an order of magnitude more water than ordinary storm tracts. As an example, a strong atmospheric river arrived in the Puget Sound region on November 6–7, 2006. Rain levels ranged were 25–75 cm in the Cascades and coastal mountains. The run-off filled the streams.

Neiman et al. (2011) studied flooding in four major watersheds in the Sound region. Two were on the western side of the Cascades, and two were on the Olympic peninsula. They found that atmospheric rivers were closely related to flooding.

The Northwest does not have hurricanes, which are tropical storms. However, they do have extratropical cyclones, and one in October 1962 was extremely severe. The barometric pressure was 960 mBar (equivalent to a category 3 hurricane), and wind gusts reached 170 mph. With the accelerating global warming now occurring, more anomalous weather events are likely. Relatively rare events, such as the hurricane, might become more common in the future. In any case, it is clear that powerful forces can and do affect the region, and they will continue to do that in the future.

Understanding these events is important to managing water reserves. Gershunov et al. (2019) examined 16 global climate models to determine how they predict atmospheric rivers. The five most accurate models confirm the variability in precipitation in most of the Western United States. Interestingly, they found that any increase in rainfall is almost entirely due to atmospheric rivers. However, O'Gorman (2015) points out that the effects of precipitation extremes are not simple. Some factors, such as convection effects and the duration and type of precipitation, are not well understood.

Climate change contributes another confounding factor to the mix. As the atmosphere warms due to climate change, it can hold more moisture and thus increase the amount of rain carried in atmospheric rivers. In recent years, the increase has been 7% per degree of surface temperature rise (Algarra et al., 2020). Based on these data, Algarra et al. predict an increase in the amount of rain that will make landfall during atmospheric river events. Shields and Kiehl (2016) used the Community Climate System Model to predict future atmospheric activity on the West Coast. They found that these events will increase in intensity with global warming.

LAND MOVEMENTS OTHER THAN EARTHQUAKES

EROSION

Segura and Booth (2010) examined the relationships of urbanization and rivers at 44 sites in the Puget lowlands. They found that more urbanized areas with more armoring to prevent erosion have fewer woody deposits, pools, and sediment. Traditional armoring of shorelines and rivers involves dams, dikes, levees, and more.

Traditional engineering to prevent erosion relies on various structures, but vegetation is only seen as a decoration. Menashe (2001) suggests that incorporating plants into the erosion control plan would be more effective. Conventional structures actually become weaker over time as they slowly decay. However, plants become stronger as they develop more extensive root systems. The side benefits of a natural component to the erosion-prevention plan include control of weeds, improved water quality, reduced run-off, improved wildlife habitat, and a more attractive landscape.

Puget Sound features a steep, irregular coastline, eroding bluffs, and mixed sand and gravel beaches. Erosion is common (Figure 10.1). Shipman (2010) found dramatic changes in the types of seeds and diatoms in the soil around buried tree stumps. The most efficient explanation is that an earthquake triggered subsidence of the ground that allowed it to be inundated. It also suggests a large earthquake between 1110 and 1150 years ago.

Great earthquakes have cause significant coastal subsidence. By examining the tidal wetlands stratigraphy around the Puget Sound, Nelson et al. (1996) concluded that oceanic and crustal factors involved. Oceanic factors include ocean volume, and crustal factors include tectonic uplift, sediment accumulation, and isostatic rebound after glacier recession.

Human activities also contribute. About one-third of the Puget Sound shoreline has been protected from erosion by seawalls, bulkheads, and other materials. That prevents shoreline erosion, but the armoring is not good for beaches and riparian zones. Fortunately, now there is some attention being paid to the armoring methods. Agriculture can result in erosion that adds sediment to streams and rivers, and dams alter the normal water flow.

FIGURE 10.1   Coastal erosion (Photograph courtesy of the US Geological Survey).

Landslides

Landslides cost lives and damage property. On March 22, 2014, a deep landslide occurred near Oso after three weeks of heavy rain. Tragically, 43 people were killed and more injured. It was the deadliest landslide in US history.

The Puget Sound's steep hills and wet winters make it particular prone to landslides. Slide-prone areas include 8.4% of the Seattle. Most of the landslides are shallow (2–3 m deep) and fast moving. Deeper movements are usually slow moving, but they can be very destructive. Rain definitely increases the likelihood of a landslide, and the bad weather can complicate the response of emergency services. Climate change will bring more extreme weather events, and landslides are sure to follow those heavy rains.

Human activities, such as development and agriculture, disrupt the natural lay of the land and make it more susceptible to landslides. Land cleared for crops encourages erosion that fills rivers and streams. In more recent times, deforestation and urban development resulted in more erosion, and dams hindered the natural movement of material (Voosen, 2020). Because landslides are often associated with extreme weather events, Cordeira et al. (2019) compared landslides from 1871 to 2012 with records of Pacific winter storms and atmospheric river events. They found that 76% of the landslides occurred during storms and 82% occurred during an atmospheric river.

Slow-moving landslides are also a problem. They rarely cost lives, but they can be extremely destructive to structures and infrastructure. The slides depend on the soil, climate, and earthquake activity of the area, but the mechanisms that initiate and maintain them are not well understood. They often occur in soils rich in clay and rock that are mechanically weak and have high levels of seasonal precipitation. Lacroix et al. (2020) reviewed the forces that control slow land-slides. Those include the overall geology, climate, and tectonics and also pre-cipitation and groundwater, earthquakes, river erosion, anthropogenic activities, and external material supply. Finnegan et al. (2019) examined the several effects on the blockage of rivers by landslides and found that wider rivers are less affected by landslides. More narrow streams can be completely blocked by the landslide. Also, rivers vary in their ability to mobilize the material in the land-slide. Land that has been moving slowly can suddenly fail with serious im-plications for people and property. Increasing pore pressure by fluids in the soil increase creep rates and can reach a stochastic point at which the system fails (Agliardi et al., 2020).

UNDERWATER LANDSLIDES

Underwater landslides can occur in the Puget Sound or on Lake Washington, which lies just to the east of Seattle. Earthquakes cause landslides, and Sherrod (2001) examined ancient forests and marshlands at four coastal locations, and showed that an earthquake triggered a landslide that cause a tsunami on the Sound. In addition, Karlin et al. (2004) examined the history of landslides in Lake Washington. From these, one can make some assumptions about the likelihood of similar events in the future. Massive landslides and slope failures resulted from large earthquakes more than once in the last 11,000 years. Similar earthquakes could happen again.

CLIMATE CRISIS

The climate crisis will have multiple significant effects on the Puget Sound region. Fires will become more common. Storms will become more severe. Sea-levels will rise to inundate low-lying areas. Summer days are likely to be warmer in years to come. Seattle usually experiences only a few days with temperatures over 90°F. By 2100, there will probably be more than two weeks each year of greater than 90°F heat.

As temperatures rise, some animals will migrate to more favorable areas. Some plants and animals will not be able to adapt and will be lost. The effects are already being seen in plants and animals. Every group is affects, but some more than others. Nearly a third of all amphibians are at risk today (Wake and Vredenburg, 2008). Birds are threatened. Since 1970, we have lost 3 billion birds or 29% of all birds (Rosenberg et al., 2019). Nearly 40% of all North American freshwater and anadromous fish are at risk (Jelks et al., 2008). Invertebrates account for 97% of all species. Bees are critical for the pollination of a large majority of human food crops, but they have been hard hit in recent years (Klein et al., 2018). Many plants are also under pressure, and they cannot easily migrate to more suitable environments (Tilman et al., 1994).

The climate has changed many times throughout the history of the Earth. Periods of hotter and colder temperatures have alternated, and the sea levels have risen and fallen. Our ancestors survived the last Ice Age. But now, it is human activities that are accelerating global warming. In the mid-19th century, humans began to use fossil fuels as never before. Now large amounts of carbon dioxide have been released into the atmosphere. Along with other greenhouse gases, carbon dioxide traps the sun's energy. The surface albedo decreases and even more heat is trapped to create a positive feedback mechanism that continues to accelerate the cycle.

Climate change has begun in the Puget Sound, and it will affect the forests in several ways. Littell et al. (2010) speculated on those effects in key species. Douglas firs (*Pseudotsuga menziesii*) will continue to grow with sufficient water, but will feel the effects of water shortages. The climate in some parts of the interior Columbia Basin and eastern Cascades will likely leave pines more susceptible to mountain pine beetles, which will become a much greater problem. Many more areas will be threatened by fire.

SEA-LEVEL RISE

One manifestation of climate change is sea-level rise that will result from the melting of the polar and glacial ice. The Rahmstorf (2007) semiempirical method is often used to estimate sea-level rise (Rahmstorf, 2007). It links sea-level rise to global mean temperatures. Kopp et al. (2014) predicts that sea levels will rise by as much as 1.2 m between 2000 and 2100. Other scientists estimate that, by 2200, our atmosphere will contain higher levels of carbon dioxide than any time over the last 650,000 years. The warming will accelerate the melting of the polar ice caps and cause the levels of the oceans to rise even faster.

The Puget Sound will not be spared from these changes. In the last century, sea levels have risen about 20 cm. Estimates now are that the average sea-level rise for Seattle will be 18–21 cm by 2050, 50–60 cm by 2100, and 75–100 cm by 2150 (Miller et al., 2018). Although many areas will be underwater with that much rise, those estimates actually understate the problem. Tidal flooding, storm surges, and the "king tides" that occur each year will put many more areas at risk (Seattle, 2021).

Coastal communities around the Sound will be more and more threatened by flooding by storm and tidal surge. Many low-lying areas contain valuable housing and infrastructure, including the airport and port facilities. The potential disruption to communities and industries is considerable and will require significant planning by governments. Barriers (e.g., dikes, levees, and sea walls) can protect some structures, but others may have move. The cost will be staggering. Governmental agencies will have to rethink development in low-lying areas, and insurance companies will need to re-evaluate the risk for development.

Many nature areas will also be affected. Takekawa et al. (2013) examined tidal marshes in the San Francisco Bay. Up to 95% of the marsh area will be inundated, and its plant communities will be lost. Those tidal wetlands are critical to the overall health of the region They filter water, slow floods, protect infrastructure, and sequester carbon. In the natural world, those plants might migrate with the rising water, but most current marsh areas are bounded by developed areas and so migration is not possible. The results are very likely to be similar in the Puget Sound region.

WILDFIRES

In September 2020, wildfires consumed more than 1335 km^2 in Washington in 24 hours. Most of the burning was in the eastern part of the state, but the Puget Sound region had its share of wildfires. Furthermore, the incidence of wildfires in the Puget Sound region is likely to increase as global temperatures rise (Barbero et al., 2015). Many of the forests in the region are quite stable. They can survive less than optimal conditions for extended periods. Halofsky et al. (2018) refer to this as "landscape inertia." The region is already experiencing reduced pre-cipitation (Abatzoglou et al., 2014) and higher temperatures.

Halofsky et al. (2018) modeled the effects of climate change and fire on the forests in western Washington through the end of the 21st century. They found that percentages of some species may change. For example, Douglas fir tolerates fire and drought better than western hemlock and western red cedar, and so, the firs may become the dominant species at lower elevations, in the Puget lowlands and northeastern side of the Olympic Mountains. This was also the case during drier times in prehistoric forests. Also, species will expand or contract their boundaries. Western hemlock will extend their range farther up the mountains.

Smoke from wildfires has become a hazard in the region (Figure 10.2). In 2018, wildfire smoke resulted in 24 days of poor air quality. Nine of those days were unhealthy for sensitive groups or unhealthy for everyone. Zauscher et al. (2013) examined the effect of fires on the local air quality. They found that 84% of the 120–400-nm particles were biomass-burning aerosols. Those particles absorb solar radiation and serve as nuclei for cloud condensation. They also are associated with adverse health effects. The make-up of the particles also varies according to the temperature of the fire. Hotter fires produce particles with more inorganic material and more soot. Cooler smoldering fires produce more organic carbon-rich aerosols. Lassman et al. (2017) studied the methods of measuring the exposure of people to smoke, a more difficult task than might be assumed due to the rapid changes in smoke concentrations. Gan et al. (2017) examined health outcomes and population exposure to smoke from wildfires in Washington 2012. They compared hospital admissions sorted by ZIP code to smoke levels de-termined by different methods. The association of smoke and health effects depended on the method used.

Land use planning and fire abatement measures can help in many cases (Syphard et al., 2013). However, that is expensive and often is in conflict with the desire of many to enjoy the benefits of living close to nature. However, the climate crisis and global warming, along with higher temperatures and stronger

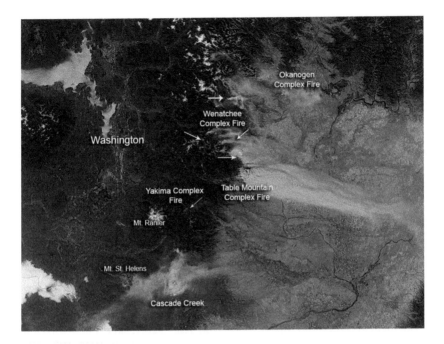

FIGURE 10.2 Wildfires in the Puget Sound (Photograph courtesy of the National Aeronautics and Space Administration).

winds, are likely to further increase the number and intensity of wildfires and the threat to humans and their property.

WINDS

The Olympic Mountains and Peninsula provides protection from the strongest winds for Puget Sound. The Sound is relatively sheltered, compared to the coast. However, the Seattle-Tacoma area can still receive sustained winds of 60–70 mph and gusts up to 90 mph. Tornadoes and hurricanes are rare. Cyclones and tropical storms are more common. Climate change will likely bring more severe wind storms to the area in the future. The damage usually involves downed trees and a degree of erosion to exposed areas.

VOLCANIC ERUPTIONS

Although we tend to forget that the Cascade Mountains are mostly volcanos, the subduction of the Juan de Fuca plate continues and is providing the energy to those volcanos. Three significant volcanoes surround the Puget Sound region: Mount Rainier, Mount Baker, and Glacier Peak.

Mount Rainier is a beautiful mountain, but it is also an active volcano (USGS, nd). It has erupted numerous times in the last 5000 years and certainly will again at some point. The exact time is not known, but the sequence of likely events is. Swarms of small earthquakes will occur below the mountain. Gas emissions and temperature will increase. The eruption itself will begin with steam and ash explosions at the summit. There may be lava or pyroclastic flows. Lahars are a serious threat. Large amounts of ash and dust could be released that would interfere with air traffic.

Volcanic eruptions are generally assumed to involve the flow of lava. Pyroclastic flows are also very dangerous. However, these are probably not the greatest threat from an eruption at Mount Rainier. The greatest threat is from lahars (Andrews, 2018). The glaciers on the mountain contain enormous amounts of ice that can be thawed rapidly. The valleys below the volcano feature steep slopes to channel the flows. Several suburbs of Seattle and Tacoma are right in the path of the flows. Those communities have over 80,000 people. Lahars travel at high speeds. The flow has the consistency of wet cement, and it takes everything in its path. Past lahars were over 150 m high. Japan and Indonesia have dammed the rivers below similar volcanoes, and those dams are effective in mitigating the flows from lahars. One river below Mount Rainier, the Puyallup River is not dammed.

The two other volcanoes in the region are also threats. Mount Baker just east of Bellingham, WA, last erupted about 6700 years ago. That event involved the collapse of the flanks, lahars, and tephra across wide areas. In 1975–1976, there was some activity but no eruption at the Sherman Crater. Glacier Peak sits in a more remote area and tends to be overlooked. However, Glacier Peak has experienced large explosive eruptions in the past 15,000 years. In 1980, Mount St. Helens gave a clear demonstration that the Cascade volcanoes are still active and can cause widespread destruction. The ash from Mount St. Helens carried over a very large area and disrupted air travel. The initial explosion removed 400 m from the top of the mountain. Any of the other Cascade volcanoes could do the same thing and certainly will at some point in the future.

The Cascade volcanoes included more than 3000 at one time or another. Most are eroded down to small hills now. But since so many have come and gone, is it possible that a new volcano could emerge? Possible, but very unlikely says Jon J. Major, the director of the Cascade Volcano Observatory (personal communication). The Juan de Fuca plate that fuels the Cascade volcanism is subducting at a fairly steady angle, and so, the melting of rock caused by the subduction has stayed in the same area over the past few hundred thousand years or more. The resulting magma has worked its way to the surface mostly along established pipelines. Thus, the current magma vents provide an existing open outlet, and there is no need for additional vents.

A New Ice Age?

The last Ice Age dramatically changed the Puget Sound and much of North America. It ended about 10,000 years ago, but interestingly, it did not end. The Pleistocene Ice Age, as it is called, has been going on for 2.5 million years and possibly much longer. Ice Ages occur about every 200 million years and last for some tens of millions of years. The Earth has had five large ice ages, and each consisted of cold (glacial) times that last tens of thousands of years and more temperature times (interglacial periods) that are only a few thousand years. More recently, ice has covered much of the Earth about every 100,000 years.

We are in one of the interglacial periods. About 20,000 years ago during the last Ice Age, an ice sheet reached across much of North America. Much of the Earth's water was locked up, and the sea levels were considerably lower. The continents reached many kilometers past the current coasts. As described in Chapter 6, the sea level was so low that a land bridge formed between Siberia and Alaska and humans migrated across that bridge to populate North America.

The cause of ice ages is not clear, but three main factors seem to be involved. First, how tilted is the Earth relative to the sun? Second, how much does the Earth wobble as it spins on its axis? Finally, what is the shape of the Earth's orbit around the sun (i.e., is it circular or oval)? These factors determine the amount of sunlight that reaches the Earth's surface, and that determines how much ice will build up at the poles.

Several theories have been suggested to explain ice ages. One of the most intriguing one comes from the great Serbian scientist Milutin Milanković and it involves the factors noted above. The Earth's orbit periodically changes from a circle to an ellipse, and the Earth tilts a bit more. These small changes affect the amount of sunlight that hits the Earths. The changes are small, but more than enough to cause the temperature variations that result in ice ages. In fact, their regularity can be used to predict the timing of the next Ice Age (Hayes et al. 1976). That was expected to begin in about 1500 years. Other factors may have a role. Changes in the amount of carbon dioxide in the atmosphere may contribute, and changes to the arrangement of the continents may influence the circulation of the oceans or atmosphere.

Another possibility is a "little ice age" (WHOI, 2021). The sun also varies in its brightness. During a grand solar minimum, the total solar irradiance is reduced by 0.25%. That seems like a very small difference, but one of these, called the Maunder Minimum, cooled the Earth in 1650–1700. The connection is only theoretical. However, temperatures on Earth during that period were much cooler than today. Some scientists predict that such an event will occur between 2020 and 2070. Current scientific thought seems to indicate that this event will slow but not fully correct the influence of human activities on global warming.

Other theories look to the Earth for the origins of the ice ages. One theory suggests that new rock was exposed when the mountain ranges were pushed up by plate tectonics (Raymo et al., 1988). That new rock weathered and eroded and fell into the oceans, taking carbon with it. Marine life used that carbon to develop calcium-carbonate shells that sequestered it from the atmosphere.

In summary, it is quite likely that the Earth will experience another ice age at some point. Of course, the big question is when. The natural factors mentioned here may determine that. But humans are a new and complicating factor. We are pumping ever-larger amounts of greenhouse gases into the atmosphere (Maslin, 2016, 2020). William F. Ruddiman (2003) asserts that a new ice age should have started a few thousand years ago, but it was prevented by human activities. Human activities, such as agricultural practices, deforestation, and biomass-burning, may have raised the levels of greenhouse gases in the atmosphere (i.e., carbon dioxide and methane) sufficiently to short-circuit the cooling and stopping the ice age. There is some evidence from ice cores that supports this theory.

Global temperatures are now rising at alarming rates. Could those rises be enough to offset the cooling that might naturally occur and postpone the next Ice Age? Some scientists believe that the next Ice Age has been delayed by 50,000–100,000 years. The answer is not clear (NASA, 2020).

Whatever the answer to that question, there is little doubt that there will be another ice age at some point. A mere 100,000 years is the blink of the eye to geology. When that happens, glaciers will again cover much of North America. They will grind their way down from the north and erase much of the geomorphology that now exists. And they will eventually recede, leaving a scoured surface behind them that will be again subjected to massive amounts of water from the melting ice that will add to the erosion.

TECTONIC PLATE MOVEMENTS

The West Coast is part of the "ring of fire" that circles the Pacific Ocean. Earthquakes tend to occur on major faults and rifts that form the ring, and they are surprisingly common. Becker and Geschwind (2016) posted a dramatic demonstration of the major earthquakes around the world from 2001 to 2015. In their work, the major faults and rifts are clearly illustrated by the number of earthquakes on them. The video provides a far better appreciation of the movement along those faults than any words.

The Puget Sound region is still involved in the subduction of the Juan de Fuca plate under the North American plate. That is likely to continue for millions of years, and the region features numerous active faults. These will continue to spawn earthquakes, volcanic eruptions, uplifts, folding, subsidence, and other geomorphic changes.

The greatest factor in the future of the region is the Juan de Fuca plate. That plate is the last remnant of the much larger Farallon plate, which has been almost completely subducted under the North American plate. Today it is positioned between two massive plates, and their movements continue to force it to subduct under the North American plate. But at some point, the subduction will end. Hawley and Allen (2019) suggest two eventual outcomes. If the subduction ends before the ridge gets to the trench, the remaining fragments of the plate will become attached to either the North American or the Pacific plate. Examples of both outcomes have been seen. The Magdalena, Guadalupe, and Monterey microplates became attached to the Pacific plate. The Llano and the Cheyenne slabs accreted onto the North American plate.

The exact fate of the Juan de Fuca plate is not understood. Its demise may result from a tear in a part of the plate that has already been subducted under the North American plate. Scientists used to believe that subduction was a smooth process. One plate slips gently under another. That is not the case at all. As a plate subducts it can warp, deform, and break, and those actions can distort the surface. All of those actions will cause significant changes to the Puget Sound region over the next millions of years.

All of these changes will be part of the larger movements of all of the plates. The plates are continuing to move. The Atlantic plate will begin to be subducted under the eastern part of the North American plate. Over 250 million years, the Atlantic Ocean will become dramatically smaller as the continents again begin to form a single supercontinent, reminiscent of Pangea. Some scientists refer to the new supercontinent as "Pangea Ultima."

Long before that time, the west coast of North America will have changed dramatically. In 10 million years, Los Angeles will have moved to the position of modern-day San Francisco. In another 50 million years, it will be near Alaska. How that will happen is not clear, but there are some theories.

All of the Puget Sound region is riding on the North American plate. Once the Juan de Fuca plate is completely subducted or attached to one or the other of the Pacific or North American plate, those two massive plates will interact directly as they do farther south below what is now the Mendocino Triple Junction. South of that point, the two plates abut one another at the San Andreas fault system. This continental transform fault currently runs about 1300 km from Southern California the Salton Sea to about Eureka in Northern California. The Pacific plate is moving northwest and continues to grind against the North American plate, which is moving southwest. In some areas, the sliding is not so smooth. The two masses can catch on each other until the stress is suddenly released as an earthquake.

The San Andreas fault divides California so that a significant part is riding on the Pacific plate. For millions of years, that part of California coast has been slipping northward. Geologists have made some interesting predictions about the future of the West Coast. These predictions are highly speculative, especially those deep into the future. For example, in 1970, Robert S. Dietz of the US Geological Survey published his predictions in an intriguing article in *Scientific American*. As noted above, he predicted that, in 10 million years, Los Angeles will be repositioned to the current San Francisco Bay Area and, in 50 million years, it will be part of Alaska. The Puget Sound region would be dramatically affected by the movement of these land masses. Christopher R. Scotese, a geologist at the University of Texas, Arlington, calculated the results of future movements on the West Coast. His fascinating website (www.scotese.com) contains projections of the Earth over the next 250 million years and generally agrees with Dietz.

So, part of California is moving north. How might that happen? The San Andreas fault is an obvious candidate for a role in this process, but there are other weak points in the North American plate that should be considered. Interestingly, those factors are active at the Southern end of the San Andreas fault near the Salton Sea. The Salton Trough is a graben or a basin created by two parallel faults. The graben becomes depressed as the land on the two outsides rises. It resulted from the stretching as the San Andreas fault and the East Pacific Rise moved. The San Andreas and the Gulf of California Rift Zone both end near the south end of the Salton Sea. A rift in the Gulf of California forms a highly geologically active area that is breaking off Baja California from Mexico. Eventually, that rift will open deeper into Mexico and then California to form an inland sea.

McCrory et al. (2009) created a model of the subduction margins in Southern California and the slab windows that have opened in that area. Volcanic events resulted from those slab windows 28.5, 19, 12.5, and 10 million years ago. With these observations, the researchers developed a model to describe the evolution of the continental margin, and they ran the model to see what it would reveal about the future of the west coast.

However, the process disclosed a problem. At about two million years, a bend in the San Andreas fault in Southern California hinders the movement. The Peninsula Ranges and the Sierras have been converging since 12.5 million years ago, but by 2 million years in the future, they will collide directly, and that pressure must be relieved somehow. McCrory et al. (2009) suggested that this stress could be relieved by a couple of possibilities. First, new faults might break through the lower Sierra Madre. Second, the movement between the plates might shift from the San Andreas to the Walker Lane system.

The Walker Lane scenario is expecially intriguing. In the next eight to ten million years, the existing rift in the Gulf of California might spread northward into the Walker Lane and eventually up to Lake Tahoe. Walker Lane is on the California-Nevada border and runs north to Death Valley and on to Mount Lassen. In the south, it connects to the San Andreas fault system (Faulds et al., 2005). With the long curve on the San Andreas fault hindering movement, the weakest point is the more natural path up Walker Lane. It could be that the San Andreas fault will become much less active as the movement and stress are assumed by this new system. In this theory, the rift that formed the Gulf of California will continue straight north on a path that includes the Salton Sea, Mono Lake, Lake Tahoe, and Pyramid Lake and a number of volcanoes. Wesnousky (2005) published an extensive study of the many faults in the Walker Lane. The westward bend of the San Andreas fault in Southern California limits movement on that fault system. The faults in the Walker Lane are far more linear and might offer a more optimal path to accommodate movement of the two plates.

The evidence supporting the Walker Lane hypothesis is growing. First, the fault system in that area already accounts for 15–25% of the movement along the North American and Pacific plates. This is a surprising amount, considering that the San Andreas fault had been assumed to be the dominant interface between the plates. Second, Baja California separated from the North American plate about seven million years ago to form the Gulf of California. A chain of volcanoes warmed the continental crust, causing it to soften and creating a series of weak spots that allowed the land to separate. Third, about 13,000 years ago, Pyramid Lake was part of an inland sea called Lake Lahontan. The lake is near a number of newly discovered faults, including the Pyramid Lake Fault, the Honey

Lake Fault, and the Warm Springs Valley Fault. These developing faults might form part of the northern part of the new rift. The rift in the Walker Lane provides an excellent opportunity to study a rift on land (Putirka and Busby, 2011; Busby, 2013). The fault system extends far beyond Lake Tahoe in the north. Interestingly, the Gorda, North American, and Pacific plates form a triple junction at Cape Mendocino in Northern California. As the Farallon plate was subducted under the North American plate, it broke into pieces. One of those pieces is the Gorda plate. The rift in Walker Lane has moved northward as the Mendocino triple junction has also moved northward. Ultimately, the rift might break towards the triple junction.

There are other possibilities. Meldahl (2015) agreed that movement on the San Andreas fault or Walker Lane might eventually affect the Puget Sound region. He also added a third in which the split occurs further east than Walker Lane. However, the end result is that, in all three, the Pacific plate continues to slip northward along the North American plate. The difference in the three futures is in where the major boundary is between the plates. In the first, more traditional view, the Pacific plate and that part of California on that plate continue moving northward. In the second, the movement is transferred from the San Andreas fault to Walker Lane as that rift continues to extend northward. In the third, a greater portion of California moves northward. In any case, the Puget Sound region will be unrecognizable.

THE FUTURE OF THE PUGET SOUND

One might conclude from this chapter that the future of the Puget Sound region is bleak. Certainly, challenges await the region. Two are probably the most serious. Tectonic movements are impossible to prevent. Those changes will happen, and the region will be changed beyond recognition. However, those changes are incredibly slow and, thus, not anything to worry about. Of course, a devastating earthquake or volcanic eruption could occur at any time. The results could be profound with many deaths and billions of dollars in damages. The real threat is climate change. These days, many things are labeled as an existential threat. Climate change truly is an existential threat. And it will manifest as many crises: extreme rain and drought, winds, and wildfires. These will put a great deal of pressure on existing systems for water, food, and land as sea-level rise. All of these will require significant amounts of planning and policy from all levels of government.

REFERENCES

Abatzoglou JT, Rupp DE, Mote PW (2014) Seasonal climate variability and change in the Pacific Northwest of the United States. *Journal of Climate* 27(5): 2125–2142.

Agliardi F, Scuderi MM, Fusi N, Collettini C (2020) Slow-to-fast transition of giant creeping rockslides modulated by undrained loading in basal shear zones. *Nature Communications* 11: 1352.

Algarra I, Nieto R, Ramos AM, Eiras-Barça J, Trigo RM, Gimeno L (2020) Significant increase of global anomalous moisture uptake feeding landfalling Atmospheric Rivers. *Nature Communications* 11: 1–7.

Allentoft ME, O'Brien J (2010) Global amphibian declines, loss of genetic diversity and fitness: A review. *Diversity* 2: 47–71.

Andrews RG (2018) *This May Be the Most Dangerous U.S. Volcano*. National Geographic. Retrieved from: https://www.nationalgeographic.com/science/article/news-most-dangerous-volcano-mount-rainier-supervolcanoes-yellowstone. April 7, 2021.

Barbero R, Abatzoglou JT, Larkin NK, Kolden CA, Stocks B (2015) Climate change presents increased potential for very large fires in the contiguous United States. *International Journal of Wildland Fire* 24: 892.

Becker N, Geschwind LR (2016) Earthquakes—2001–2015. Science on a Sphere. National Oceanic and Atmospheric Administration. http://sos.noaa.gov/Datasets/dataset.php?id=643

Busby CJ (2013) Birth of a plate boundary at ca. 12 Ma in the Ancestral Cascades arc, Walker Lane belt of California and Nevada. *Geosphere* 9: 1147–1160.

Cordeira JM, Stock J, Dettinger MD, Young AM, Kalansky JF, Ralph FM (2019) A 142-year climatology of Northern California landslides and atmospheric rivers. *Bulletin of the American Meteorological Society* 100: 1499–1509.

Dunwiddie PW, Alverson ER, Martin RA, Gilbert R (2014) Annual species in native prairies of South Puget Sound, Washington. *Northwest Science* 88: 94–105.

Encyclopedia (nd) *Invasive Species*. Encyclopedia of the Puget Sound. Retrieved from: https://www.eopugetsound.org/terms/174. April 6, 2021.

Faulds JE, Henry CD, Hinz NH (2005) Kinematics of the northern Walker Land: An incipient transform fault along the Pacific–North American plate boundary. *Geology* 33: 505–508.

Feist BE, Buhle ER, Baldwin DH, Spromberg JA, Damm SE, Davis JW, Scholz NL (2018) Roads to ruin: conservation threats to a sentinel species across an urban gradient. *Ecological Applications* 27: 2382–2396.

Finnegan NJ, Broudy KN, Nereson AL, Roering JJ, Handwerger AL, Gennett G (2019) River channel width controls blocking by slow-moving landslides in California's Franciscan mélange. *Earth Surface Dynamics* 7: 879–894.

Gan RW, Ford B, Lassman L, Pfister G, Vaidyanathan A, Fischer E, Volckens J, Pierce JR, Magzamen S (2017) Comparison of wildfire smoke estimation methods and associations with cardiopulmonary-related hospital admissions. *GeoHealth* 1: 122–136.

Gershunov A, Shulgina T, Clemesha RES, Guirguis K, Pierce DW, Dettinger MD, Lavers DA, Cayan DR, Polade SD, Kalansky J, Ralph FM (2019) Precipitation regime change in Western North America: The role of atmospheric rivers. *Scientific Reports* 9: 9944.

Hallmann CA, Sorg M, Jongejans E, Siepel H, Hofland N, Schwan H, et al. (2017) More than 75 percent decline over 27 years in total flying insect biomass in protected areas. *PLoS One* 12: e0185809.

Halofsky JS, Conklin DR, Donato DC, Halofsky JE, Kim JB (2018) Climate change, wildfire, and vegetation shifts in a high-inertia forest landscape: Western Washington, U.S.A. *PLoS ONE* 13(12): e0209490.

Hawley WB, Allen RM (2019) The fragmented death of the Farallon plate. *Geophysical Research Letters* 46: 7386–7394.

Hayes JD, Imbrie J, Shackleton NJ (1976) Variations in the Earth's orbit: Pacemaker of the Ice Ages. *Science* 194: 1121–1132.

Hristov P, Shumkova R, Palova N, Neov B (2020) Factors associated with honey bee colony losses: A mini-review. *Veterinary Sciences* 7: 166.

Jelks HJ, SJ Walsh, NM Burkhead, Contreras-Balderas S, Díaz-Pardo E, Hendrickson DA, Lyons J, Mandrak NE, McCormick F, Nelson JS, Platania SP, Porter BA, Renaud CB, Schmitter-Soto JJ, Taylor EB, Warren Jr ML (2008) Conservation status of imperiled North American freshwater and diaddromous fishes. *Fisheries* 33(8): 372–407.

Karlin RE, Holmes M, Abella SEB, Sylwester R (2004) Holocene landslides and a 3500-year record of Pacific Northwest earthquakes from sediments in Lake Washington. *GSA Bulletin* 116: 94–108.

Kopp RE, Horton RM, Little CM, Mitrovica JX, Oppenheimer M, Rasmussen DJ, Strauss BH, Tebaldi C (2014) Probabilistic 21st and 22nd century sea-level projections at a global network of tide-gauge sites. *Earth's Future* 2: 383–406.

Klein AM, Boreux V, Fornoff F, Mupepele AC, Pufal G (2018) Relevance of wild and managed bees for human well-being. *Current Opinion in Insect Science* 26: 82–88.

Lacroix P, Handwerger AL, Bièvre G (2020) Life and death of slow-moving landslides. *Nature Reviews Earth and Environment* 1: 404–419.

Lassman W, Ford B, Gan RW, Pfister G, Magzamen S, Fischer EV, Pierce JR (2017) Spatial and temporal estimates of population exposure to wildfire smoke during the Washington state 2012 wildfire season using blended model, satellite, and in situ data. *GeoHealth* 1: 106–121.

Littell JS, Oneil EE, McKenzie D, Hicke JA, Lutz JA, Norheim RA, Elsner MM (2010) Forest ecosystems, disturbance, and climatic change in Washington State, USA. *Climate Change* 102: 129–158.

Loss SR, Will T, Marra PP (2015) Direct mortality of birds from anthropogenic causes. *Annual Review of Ecology, Evolution, and Systematics* 46: 99–120.

Maslin M (2016) Forty years of linking orbits to ice ages. *Nature* 540: 208–209.

Maslin M (2020) Tying celestial mechanics to Earth's ice ages. *Physics Today* 73: 48–53.

Mason THE, Stephens PA, Gilbert G, Green RE, Wilson JD, Jennings K, Allen JRM, Huntley B, Howard C, Willis SG (2021) Using indices of species' potential range to inform conservation status. *Ecological Indicators* 123: 107343.

McCrory PA, Wilson DA, Stanley RG (2009) Continuing evolution of the Pacific–Juan de Fuca–North American slab window system—A trench-ridge-transform example from the Pacific Rim. *Tectonophysics* 464: 30–42.

Meldahl K (2015) *Surf, Surf, Sand, and Stone. How Waves, Earthquakes, and Other Forces Shape the Southern California Coast.* University of California Press. Retrieved from: https://doi.org/10.1525/9780520961852. October 18, 2021.

Menashe E (2001) Bio-structural erosion control: Incorporating vegetation in engineering designs to protect Puget Sound shorelines. *Puget Sound Research* Retrieved from: http://www.greenbeltconsulting.com/assets/pdfs/BioStructuralErosionControl.pdf. October 18, 2021.

Miller IM, Morgan H, Mauger G, Newton T, Weldon R, Schmidt D, Welch M, Grossman E (2018) *Projected Sea Level Rise for Washington State – A 2018 Assessment.* A collaboration of Washington Sea Grant, University of Washington Climate Impacts Group, Oregon State University, University of Washington, and US Geological Survey. Prepared for the Washington Coastal Resilience Project. Retrieved from: https://cig.uw.edu/resources/special-reports/sea-level-rise-in-washington-state-a-2018-assessment/. April 7, 2021.

NASA (2020) *There Is No Impending "Mini Ice Age."* NASA Global Climate Change. Retrieved from: https://climate.nasa.gov/blog/2953/there-is-no-impending-mini-ice-age/. April 7, 2021.

Neiman PJ, Schick LJ, Ralph FM, Hughes M, Wick GA (2011) Flooding in Western Washington: The connection to atmospheric rivers. *Journal of Hydrometerology* 12: 1337–1358.

Nelson AR, Shennan I, Long AJ (1996) Identifying coseismic subsidence in tidal-wetland stratigraphic sequences at the Cascadia subduction zone of western North America. *Journal of Geophysical Research* 101: 6115–6135.

O'Gorman PA (2015) Precipitation extremes under climate change. *Current Climate Change Reports* 1:49–59.

Putirka KD, Busby CJ (2011) Introduction: Origin and evolution of the Sierra Nevada and Walker Lane. *Geosphere* 7:1269–1272.

Rahmstorf S (2007) A semi-empirical approach to projecting future sea-level rise. *Science* 315: 368–370.

Raymo ME, Ruddiman WF, Froelich PN (1988) Influence of late Cenozoic mountain building on ocean geochemical cycles. *Geology* 16: 649–653.

Rosenberg KV, Dokter AM, Blancher PJ, Sauer JR, Smith AC, Smith PA, Stanton JC, Panjabi A, Helft L, Parr M, Marra PP (2019) Decline of the North American avifauna. *Science* 366: 120–124.

Ruddiman WF (2003) The anthropogenic greenhouse era began thousands of years ago. *Climatic Change* 61: 261–293.

Rutz JJ, Steenburgh WJ, Ralph FW (2014) Climatological characteristics of atmospheric rivers and their inland penetration over the Western United States. *Monthly Weather Review* 142: 905–921.

Seattle (2021) *Projected Climate Changes*. Seattle Public Utilities. Seattle.gov. Retrieved from: http://www.seattle.gov/utilities/protecting-our-environment/community-prog rams/climate-change/projected-changes#:~:text=Sea%2DLevel%20Rise,inches%20 during%20the%20past%20century.&text=Central%20estimates%20indicate%20 that%20Seattle,and%2047%20inches%20by%202150. April 7, 2021.

Segura C, Booth DB (2010) Effects of geomorphic setting and urbanization on wood, pools, sediment storage, and bank erosion in Puget Sound streams. *Journal of the American Water Resources Association* 46: 972–986.

Sherrod B (2001) Evidence for earthquake-induced subsidence about 1100 yr ago in coastal marshes of southern Puget Sound, Washington. *Geology* 113: 1299–1311.

Shields CA, Kiehl JT (2016) Atmospheric river landfall-latitude changes in future climate simulations. *Geophysical Research Letters* 43: 8775–8782.

Shipman H (2010) The geomorphic setting of Puget Sound: Implications for shoreline erosion and the impacts of erosion control structures. In: Shipman H, Dethier MN, Gelfenbaum G, Fresh KL, Dinicola RS (editors). *Puget Sound Shorelines and the Impacts of Armoring—Proceedings of a State of the Science Workshop, May 2009*, pp. 19–34. U.S. Geological Survey, Washington.

SOEM (nd) *Earthquakes*. Seattle Office of Emergency Management. Retrieved from: https://www.seattle.gov/Documents/Departments/Emergency/PlansOEM/SHIVA/2 014-04-23_Earthquakes(0).pdf. April 6, 2021.

Syphard AD, Bar Massada A, Butsic V, Keeley JE (2013) Land use planning and wildfire: Development policies influence future probability of housing loss. *PLoS ONE* 8(8): e71708.

Takekawa JY, Throne KM, Buffington KJ, Spragens KA, Swanson KM, Drexler JZ, Schoellhamer DH, Overton CT, Casazza ML (2013) *Final Report for Sea-Level Rise Response Modeling for San Francisco Bay Estuary Tidal Marshes*. Open-File Report 2013-1081. US Geologic Survey. Retrieved from: https://pubs.er.usgs.gov/ publication/ofr20131081.

Tilman D, May R, Lehman CL, Nowak MA (1994) Habitat destruction and the extinction debt. *Nature* 371: 65–66.

USGS (nd) *Future Eruptions at Mount Rainier*. US Geologic Survey. Retrieved from: https://www.usgs.gov/volcanoes/mount-rainier/future-eruptions-mount-rainier#:~:text=Mount%20Rainier%20is%20behaving%20about,the%20shore%20along%20Commencement%20Bay. April 7, 2021.

Voosen P (2020) A muddy legacy. *Science* 369: 898–901.

Wake DB, Vredenburg VT (2008) Are we in the midst of the sixth mass extinction? A view from the world of amphibians. *Proceedings of the National Academy of Sciences of the United States of America* 105: 11466–11473.

Wesnousky SG (2005) Active faulting in the Walker Lane. *Tectonics* 24: TC3009.

WHOI (2021) *Are We on the Brink of a 'New Little Ice Age?'* Woods Hole Oceanographic Institution. Retrieved from: https://www.whoi.edu/know-your-ocean/ocean-topics/climate-ocean/abrupt-climate-change/are-we-on-the-brink-of-a-new-little-ice-age/. April 7, 2021.

Zauscher M, Wang Y, Moore M, Gaston C, Prather K (2013) Air quality impact and physicochemical aging of biomass burning aerosols during the 2007 San Diego wildfires. *Environmental Science & Technology* 47. 10.1021/es4004137.

Index

Milton Keynes UK
Ingram Content Group UK Ltd.
UKHW022048141024
449569UK00023B/846

9 781032 201184